NOV 1 8 2004

Petrolia

The Johns Hopkins University Press • *Baltimore and London*

Creating the North American Landscape

Gregory Conniff   Bonnie Loyd
Edward K. Muller   David Schuyler
Consulting Editors

George F. Thompson
Series Founder and Director

Published in cooperation with
the Center for American Places,
Santa Fe, New Mexico, and
Harrisonburg, Virginia

PETROLIA

THE LANDSCAPE OF AMERICA'S
FIRST OIL BOOM

Brian Black

Unless otherwise indicated, all maps and
illustrations are from the Pennsylvania
Historical and Museum Commission,
Drake Well Museum Collection, Titusville,
Pennsylvania.

Title page photo: Pioneer Run Wells, 1866

9 8 7 6 5 4 3 2 1

The Johns Hopkins University Press
2715 North Charles Street
Baltimore, Maryland 21218-4363
www.press.jhu.edu

Library of Congress Cataloging-in-
Publication Data will be found at the
end of this book.
A catalog record for this book is available
from the British Library.

ISBN 0-8018-6317-1

For Christina

Contents

Acknowledgments

All history writing seeks to fill existing gaps. In the case of early oil, many books have been written to tell the exciting stories of boom and bust, leaving few gaps in the story. For this and other reasons, a new book has not been written specifically on Pennsylvania's boom—that is, the world's first oil boom—since the 1940s. As I set out to update this story, I located gaps, not in the detail of earlier work, but in logic. As I seek to ask new questions of past events in this place, my greatest debts may be intellectual.

As I considered examples of the work of earlier historians, I found some that possessed portions of a new perception that had yet gone unstated. Most importantly, the writing and spirit of Ida Tarbell rose like a beacon guiding me beyond the romance and riches to the human and natural story available in the oil country of Pennsylvania. With my destination realized, the writing of William Culp Darrah, a true renaissance scholar of botany, history, and the natural sciences, offered an approach through his fascinating analysis of development and decline in Pithole, Pennsylvania. Even with this spirit and formula for analysis, *Petrolia* would have been impossible without the photographic work of John Mather. His grand catalogue of this boom stands as a symbol of why preservation is important: first, I thank him for believing that this place warranted his attention; second, I express appreciation to the countless individuals who have maintained the collection's safety and organization over one hundred years. With glass-print negatives, this security is no small feat. This hard work, in many ways, preserved not only a photo collection but also a place. My work is made possible by their efforts.

The debts that have grown out of this project have been acquired in Lawrence, Kansas, New York, New York, many stops in Pennsylvania, and a few other out-of-the-way corners of the United States. This project grew out of the introspection and thought encouraged by the American Studies program at the University of Kansas. It was my privilege to help to coordinate a wonderful community of students and professors brought together by the Rockefeller Foundation's monthly colloquia at the University of Kansas on Nature, Technology, and Culture. Through such work I came into contact with thinkers and scholars who have contributed greatly to this work, including Paul Sutter, Paul Hirt, Mike French, Steve Hamburg, Paul Rich, Leos Jelecek, Amanda Rees, Sterling Evans, and many others. Most importantly, however, this project grows out of the scholarship, teaching, and person of Donald Worster. A creative and intense scholar, Don fostered my study of Petrolia through his thought and maybe more importantly through his generosity of time, spirit, and intellect. Although hidden away in the remote Kansas prairie, Don and Bev Worster have been found by many fortunate students. I treasure that I could be one.

John G. Clark, Dennis Domer, and David M. Katzman gave considerably of their time and intellect. I am also indebted to Norm Yetman, Angel Kwollek-Folland, James Shortridge, Peter Mancall, Barry Shank, Burt Perretti, Ann Schofield, Bryan LeBeau, Richard Horowitz, Bill Cronon, Patty Limerick, Richard Francoviglia, David Schuyler, David Nye, Peter Stitt, Michael Birkner, Magadalena Sanchez, Bill Bowman, John Stilgoe, Gabor Boritt, Norman Forness, Katie Clay, Helen Schumaker, Dan Ricci, Max McElwain, Stuart Tarr, Jim Eber, Paul Baker, Donald White, John Boland, and others. Spiritual and intellectual debts were also incurred by basketball, guitar, or sailboat in association with Adam Rome, James Pritchard, and Diane Debinski.

The research that went into this book involved generous support of one type or another from the staffs of the Hall Center for the Humanities, American Studies department, and libraries of the University of Kansas; the Pennsylvania History and Museums Commission (PHMC); the Mid-America American Studies Association; the National Endowment for the Humanities; the Rockefeller Foundation; Christopher Chadbourne and Associates; the Oil Heritage Corporation; and particularly the Drake Well Museum and Archive in Titusville, Pennsylvania, with special appreciation to Barbara Zolli, Anne Stewart, David Weber, and Susan Beates. The production of this book owes a great deal to George Thompson and the staff of editors and readers working with the Center for American Places. I thank the center and Skidmore College for support and encouragement in bringing *Petrolia* to print. Additionally, I offer gratitude to the editorial staffs and readers of the journals in which portions of this book were first published: *Landscape, Pennsylvania History*, and *Environmental History*.

Finally, I come to the family who has aided me in getting through the oily portion of my academic life. Clyde and Rita have supported me mightily in all that I have done. Their inspiration, with that of Hazel, Dick, Gladys, and particularly Clyde, Sr., have buoyed my efforts and spirits in more ways than they can ever know. And Jen, Joe, Eliza, and Will, as well as each of my in-laws, have provided moral support when it was needed most. Christina, who has worn the mud and grime of Petrolia, has been the energy and passion behind my work. And now she has also brought me young Ben and Sam, to whom I look forward to introducing Pennsylvania's oil boom. I am thankful for the efforts of PHMC, oil buffs, and local oil heritage groups, who have each worked to preserve for Ben, Sam, and others the opportunity to experience a sense of this evocative place.

As one reads this manuscript, it will become obvious that more than to any of these individuals, this research and writing is dedicated in spirit to a place—a place now lost. I do not hail from the steep slopes that surround Oil Creek, but I have come to know the area and to appreciate its unique story. I came to it as an investigator or observer more than as a scholar. I have learned a great deal in the archives of the region, yet I have learned more on morning walks through the damp underbrush along Oil Creek.

In this history, I attempt a more intimate portrayal of the history than has ever before been accomplished. I attempt to tell a story that goes beyond the commodity that defined the meaning of this place and to return to it *its* meaning. I am moved to do this by my own history in such a place—a former site of industry that now struggles for an identity. In Pennsylvania, most locales have seen their heyday come and go—for some, their any day has also come and gone. I have spent my life mesmerized by the people of such locales, the human remnants of the place's cumulative character. When the industry moved on, many workers, having grown accustomed to the place, remained behind to do what they might. For most, this life never fulfilled expectations. In my travels throughout the state, I came to view their story as tragic.

Certainly, residents of company or industrial communities are beneficiaries of a living made from harvesting resources, but they are also subject to the inevitable decline of their social and natural environment. Indeed, traditionally, these earliest industrial communities have almost always been abandoned by the industries that created them. Too often a mode of production or land use moves on, and the human communities are left with nothing in a place that has become desolate or even dangerously contaminated. Such a system of ethics assumes that a place and its residents have neither meaning nor significance unless they produce the necessary resource. In the end, this book salutes the people and places that remain to tell the story when the industry has moved elsewhere. Like living echoes, they continue to resonate so that someday we can apply what they have seen and learned to upcoming situations and decisions.

Petrolia

Introduction

The Persistence of Oil on the Brain

Little do I remember of . . . the increased comforts of life or moving into the new home on the hillside above the town by this time known as Rouseville. But the change in the outlook on the world about me, I do remember. We had lived on the edge of an active oil farm and oil town. No industry of man in its early days has ever been more destructive of beauty, order, and decency than the production of petroleum. All about us rose derricks, squatted engine-houses and tanks; the earth about them was streaked and damp with the dump-ings of the pumps, which brought up regularly the sand and clay and rock through which the drill had made its way. If oil was found, if the well flowed, every tree, every shrub, every bit of grass in the vicinity was coated with black grease and left to die. Tar and oil stained everything. If the well was dry a rick-ety derrick, piles of debris, oily holes were left, for nobody ever cleaned up in those days.[1]

The Alleghenies of northwestern Pennsylvania rise and fall with striking im-mediacy. Similar to the random lumps rising in a thin blanket if the sheet be-neath has not been pulled tight, these short hills fail to reach mountain sta-tus but together unite to give the landscape a bulky and distinctly haphazard look. Mixed deciduous forest fills most of the swells before trailing off to the east to become the crop of the Allegheny National Forest. Sprinkled at the foot of these hills lay many small towns, most once built to house or profit from a bygone industry. One immediately conjures up visions of iron, steel, lumber, and other manufacturing; however, the greatest story in these hills grates on most contemporary stereotypes. These hills, and not the flatlands of Texas, the deserts of Saudia Arabia, or the waters of the Gulf of Mexico, harbored the world's first extraction of crude oil.

As the world's largest oil producer from 1859 to 1873, this valley compounded its importance and created a legacy far broader than that of its own industry. This more significant legacy did not escape the stiletto-pen of Ida Tarbell. Raised within the industry, Tarbell recognized broader patterns emanating from this site. Specifically, the oil boom epitomized a problematic national confidence of the mid- to late-nineteenth century, a belief that unlimited technological development would lead inevitably to human progress. Tarbell voiced these observations in direct contrast to the celebratory literature placing the oil boom within the nation's increasingly aggressive march toward economic prosperity.

Compare her perspective, for instance, to that of *Petroleum*, written by the Reverend S. J. M. Eaton in 1866 as the first extensive history of the world's premier oil-producing region. In his mind, divine providence had manifested itself through the geology of this mountainous region. "The [oil] field is large," he wrote, and "the source of supply is exhaustless. It has evidently been a product of earth from the beginning. It has been one of God's great gifts to his creatures, designed for their happiness; but kept locked up in his secret laboratory, and developed only in accordance with their necessities. And now in our own day, and in these ends of the earth, the great treasure house has been unlocked, the seal broken, and the supply furnished most abundantly."[2]

Muckrakers such as Tarbell burst the bubble of such idealization. Yet, like investigative journalists, they could only do so on a limited basis: proving the malice of a specific robber baron; the carelessness of a single industry, such as meat packing; the exploitation of young or immigrant workers. There could be no wholesale judgment cast against the era, for such commentary qualified as blasphemy against much of what America stood for. In the end, the vast majority of these sites, including that of the Pennsylvania oil boom, escaped extensive critical consideration.

The oil industry in general, of course, did not remain entirely unscathed. Tarbell's famous exposé of the Standard Oil Company and the business practices of John D. Rockefeller Sr. waged a war against the giant who came to sit atop the oil industry.[3] The story of one man's greed overshadowed that of the thousands of boomers who preceded him. Tarbell's analysis used Rockefeller as a representative figure in hopes of prodding Americans to question ideas of success and acceptable development. The excerpt above reveals that Tarbell sought to pierce more deeply to the nation's very ideas of economic expansion. Tarbell accused industrialists of misusing human labor; yet this paragraph refused to stop there. Tarbell attached this same brute ethic of capitalist gain to corporations' management and use of resources—including humans and the natural environment.

I think of the spirit of Tarbell's writing as I hike the young forest of the Oil Creek valley. With each step I recede from the nearest town, Titusville, which is four miles upriver. I stand roughly one hundred miles directly north of Pittsburgh and forty miles south of Erie (map I.1). Interstate 80 crosses this country about thirty miles to the south, and few tourists willfully exit at Clarion to venture the small, indirect roads to the oil regions. The largest road entering the region is Pennsylvania Route 36, which comes from the southeast. Beginning in Altoona, the route's name, the Colonel Drake Highway, identifies it as a legacy of oil's discoverer. Today, the route's tight passages give travelers ample opportunity to ponder its name as they follow

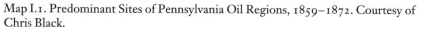

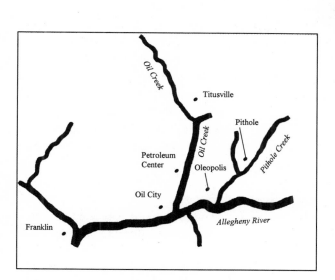

Map I.1. Predominant Sites of Pennsylvania Oil Regions, 1859–1872. Courtesy of Chris Black.

Fig. I.1. Oil Creek, 1994. Photo by the author.

the plodding trucks that carry out lumber from the vast stands of hemlock, oak, and pine still to be found near Pennsylvania's center.

The forest here holds cultural secrets as well as ecological ones. The scattered physical remains of the oil industry offer us an opportunity to recall the scene of the 1860s. This valley and its heritage can teach us a great deal about our entire industrial history. The physical squalor, as well as the abandoned equipment still littering the forest, indicates the broader ethos that guided the use of natural resources in this place. The apparatuses used for the oil industry had never before been employed. The type of extraction involved had only been practiced on water, which is entirely different in value from mineral resources. In 1859, this new industry demanded an entirely new and different method of land use. The new industry required that speculators, as well as the intrigued public, reconfigure their very view of the world, its resources, and how those resources should be managed and used.

As the forest grows up around the forgotten remnants of an industry moved elsewhere, it presents the viewer with a record of the culture of transience that dominated life during this valley's oil boom (fig. I.1). In the end, this place of the 1860s proved a harbinger of the thinking that guided American industry and land use throughout the nineteenth and twentieth centuries. Its legacy becomes each of the American industrial sites that followed, including the Hanford nuclear waste site in Washington; the Love Canal toxic dump in New York and its sister Superfund sites throughout the nation; any of the countless coal strip mines in Pennsylvania, West Virginia, Kentucky, Tennessee, Ohio, and many Western states; the forests of the Great Lakes; the agricultural fields of the Great Plains, as well as the site of the Yucca Mountain Nuclear repository in Nevada. The ethics and values exhibited in each of these contemporary scenes appeared earlier in the Oil Creek valley. They came about, at least partly, because of oil.

The Oil Creek valley's reign as the world's oil king lasted from 1859 until 1873. During those fourteen years, humans extracted fifty-six million barrels of oil from this ground, grossing investors and speculators over $17 million by 1865.[4] As the scrupulous Ida Tarbell relayed earlier, few speculators gave thought to "cleaning up," a suggestive phrase about ameliorating the incredibly wasteful and destructive methods employed by the early industry. Developers and speculators in the oil business believed that they had no reason to consider the region's future; they would take their fortune back home or on to the next frontier. This boom and its transient workforce created this place's meaning; more troubling, the popular media guided the interest that defined the region from the outside inward.

Regardless of its conception, an important story of historical change lay hidden in the place. Like any land users, the men and a very few women, while residing in this valley and orchestrating its oil boom, made decisions and choices based on a set of values and an economic order. For many of these decisions, the boomers left an artifactual record: the physical landscape, which offers contemporary viewers the opportunity to deconstruct the scene in order to re-create the values of the inhabitants (fig. I.2). If boiled down and simplified, these various values combine to form a basic approach toward natural resource use—an ethic.

The Oil Creek valley offers a uniquely confined example of the ethics that characterized this era of wondrous change. Initially, it appears that the excitement over petroleum's potential uses and the supply's seeming inexhaustibility dominated every onlooker's view of the oil industry—that boom consists of boomers and few others. Certainly, oil thrived as an industry

Fig. I.2. Great Western Run Wells, 1864

pulled up by its bootstraps with no thought to the future of the place or those living in it. Locally, however, the view becomes clouded with complications. For those reading of the boom, such recklessness made the early years terrifically exciting. For those living it, the boom also contained frustration, loneliness, and—the antithesis of boom—bust. The stark necessities of extractive development intensified this reality: while journalists, writers, and speculators agreed that the oil industry had a great future, it would not necessarily take place near Oil Creek. This valley became merely a stepping-stone—actually a launching point—from which an industry would spring and spread across the globe.

The influence of outsiders through national and international investment only served to intensify the industry's temporal nature.[5] This situation introduced a vast gulf separating company decisionmakers from the resource they were extracting. A place such as the Oil Creek valley lost its meaning as a self-defining cultural or ecological community to become a cog in the industrial production of crude oil. It led the way for a period in which Americans came to accept increasingly intensive manipulation of their natural surroundings.

Ecologists discuss the impact of human influence in such a place within the rubric "disturbance" or "modification."[6] Obviously, an oil boom contains a daunting level of change. An ecologist may abandon attempts to assess its extent and seize on something concrete, such as measuring regional pollution rates or water purity. Such analysis neglects the source of the disturbance: the culture of the place. Environmental history offers the opportunity to create a historical story containing the critical scrutiny normally reserved for the scientific perspective. In the case of Pennsylvania's oil boom, environmental history allows one to re-create one of the earliest examples of the culture of massive disturbance—the culture that remains a mainstay of American economic development. It allows history to pick up where the muckrakers left off.

This analysis requires that one question the view of progress that the Reverend Eaton and others found only worthy of celebration. Such an ethos permeated the 1860s, even resulting in the catchy lyrics of "Oil on the Brain," the 1865 oil-boom tune that swept the nation:

> The Yankees boast that they make clocks which "just beat all creation."
> They never made one could keep time with our great speculation.
> Our stocks, like clocks, go with a spring, wind up, run down again;
> But all our strikes are sure to cause "Oil on the Brain."
>
> The lawyers, doctors, hatters, clerks, industrious and lazy,
> Have put their money all in stocks, in fact, have gone "oil crazy."
> They'd better stick to briefs and pills, hot irons, ink, and pen,
> Or they will "kick the bucket" from "Oil on the Brain."[7]

The oil-boom tunes teamed with other portions of popular culture to create an image of the early oil industry focusing on immediate monetary gain. It was a most attractive image to an audience of eager individuals, each seeking his or her own personal boom. In its own way, such music aided development by acting as a type of anesthetic, numbing one's ability to discern the negligence. The history of this place tells a less glamorous story.

Historians have long distinguished the aggressiveness of industrial change before the Civil War from that coming after it. The first industrial revolution introduced a new intensity of human resource use, but it paled in comparison with that taking shape after the Civil War. Of this earlier industrial period, anthropologist Anthony F. C. Wallace wrote, "The machine was in the garden, to be sure, but it was a machine that had grown almost organically in its niche, like a mutant flower that was finding a congenial place among the rocks, displacing no one else and in fact contributing to the welfare of the whole."[8]

Wherever it left its boot print, the second, or later industrial revolution, would strain such bucolic description. Howard Mumford Jones led a number of historians pointing at the experience of war to explain the change between these two periods. Although the war did not directly cause technological innovation, Jones said, "it dramatized ingenuity, it accustomed people to mass and size and uniformity and national action, it got them used to ruthlessness, it made clear the dominant place of energy in the modern state." Energy became at once the most needed commodity and the medium for attaining it. Competition, fortune, and these new ideas of resource use dramatically altered this period of industrial change. Jones continued, "There was a continent to ravage. The discovery that energy could be channeled into vast and profitable projects of destruction created in the era a kind of fierce, adolescent joy in smashing things."[9]

The cultural drive to harvest resources and make them profitable at any cost had become widespread by the close of the nineteenth century. While industrial development had previously been a slow, plodding process, petroleum development derived from the rapidity of this emerging ethos. The oil rush followed quickly on the heels of the 1849 gold rush to California, further altering cultural ideas of progress and acceptable land use. During this era, few tales enjoyed such widespread appeal in the United States as did the one of incredible wealth realized from natural resources. The idea of the valueless becoming valuable filled every day with the possibility of locating one's fortune right beneath one's nose.

Many observers argued that Americans were particularly well poised to reap the economic benefits of this new industrial intensity. Certainly, the nation's expansive supply of natural resources represented a vast storehouse to be realized and seized. More importantly, however, Americans had a receptiveness to change included among their founding principles. Unlike other nations, the United States replaced tradition with trust in individual opportunity and development. Years earlier Frenchman Alexis de Tocqueville had noticed Americans' knack for quickly adapting their visions of progress to the times:

> Man gets accustomed to everything. He gets used to every sight. . . . [He] fells the forests and drains the marshes. . . . The wilds become villages, and the vil-

lages towns. The American, the daily witness of such wonders, does not see anything astonishing in all this. This incredible destruction, this even more surprising growth, seem to him the usual progress of things in this world. He gets accustomed to it as to the unalterable order of nature.[10]

The event of one's plague becoming another's boon did not occur daily; however, as Tocqueville noted, it happened often enough that Americans did not need to fear change.

Awesome and even fearsome changes overtook industrial development during and following the Civil War. The dangers of the young petroleum industry, for instance, threatened every speculator and regional resident. However, instead of reacting with fear or caution, Americans most often were intrigued. The presence of what should have been intimidating details in journalistic accounts led to the re-creation of the Oil Creek valley as a mythic place in the popular culture. While other aspects of oil may have been more important to economic development, the lasting impact of this cultural reconception of land use remains the least-known portion of the early industry's legacy. It possessed a broad influence over the industries that would follow.

In the terminology of historian Carolyn Merchant, what began on August 27, 1859, in this remote place made up no less than an "ecological revolution."[11] Such revolutions occur when humans press their natural environment in a new or different fashion than others have before them. If it becomes a practice that others implement elsewhere, a revolution has taken place in the way that the human and natural worlds interact. Such momentary incursions can eventually become distinct shifts in the border's location. The Oil Creek valley possesses such a distinction.

There is at least one major pitfall in re-creating this story: One must resist equating extraction with degradation. The oil boom created an industrial wasteland, but that does not emblazon it as a simple story of ruin. Writing in the first half of the twentieth century, wildlife biologist Aldo Leopold, in *A Sand County Almanac*, called for "an ecological interpretation of history." Such a history is at the root of Merchant's restructuring of our human past. In this form, human history composes only a chapter in a long chronicle of natural history; humans are a species within a larger biological world and history. Leopold offers a guide to formulating histories of communities such as the Oil Creek valley: "Many historical events, hitherto explained solely in terms of human enterprise, were actually biotic interactions between people and land. The characteristics of the land determined the facts quite as potently as the characteristics of the men who lived on it."[12]

The biotic reaction in this valley took place when the naturally occurring pockets of fossil fuels met the human enterprise and dogged pursuit of wealth that created the period of American industrialization. Some may argue that anything that humans, as a portion of nature, create or accomplish is a natural act and should therefore not be separated from the natural environment; however, when humans make the choice to harvest or reap resources beyond their needs for subsistence, they place themselves outside of all other life on Earth. This human precariously stands atop little except his own technological resourcefulness. He has begun a vicious cycle of reactionary problem solving, and his only weapon is his own ingenuity.

The early oil industry offers a model for considering the effects of massive industrialization on individual and community views of the landscape—most basically, the culture's environmental ethic.[13] During the 1860s, industry sacrificed the Oil Creek valley. An ecological, natural landscape became purely a commodity. Such a disturbance extends well beyond the immediate act of bringing oil to the surface; this process reevaluates the worth of a locale to the extent that even natural entities, such as Oil Creek, become commodities. While all mining causes the valuable resource's surroundings to be reevaluated in some fashion, oil in the 1860s possessed characteristics that distinguish it from any other previously mined resource. The most distinct characteristic, of course, is the mineral's liquid form, which significantly alters the application of laws of ownership (and the organization that goes with them). Unlike other extractive enterprises, the pursuit of oil resembles the pattern of the sly mongoose: the inside of the egg extracted but the shell left intact to deceive the observer. The remaining scene of extraction, then, serves as a record of the priorities of the industry in the 1860s.

As one stands today atop the swell beside Oil Creek and looks down on sites of a bygone era, such as Petroleum Centre, Cherrytree Run, Pithole, the Tarr family farm, and Pioneer Run, it becomes obvious that the forecast of an endless supply of oil in this valley was terribly mistaken. There are many portions of the oil industry that came to fruition in this valley (drilling techniques, railroad tank cars, pipelines, derrick design, ownership rights, offshore drilling), but the emptiness of abandonment appears to be the early oil industry's most omnipresent reminder.

As it went from this valley, the resource of petroleum became an international commodity. Today, the perpetual industrial growth with which the industry proceeds in a variety of cultures and economic systems depends on its ability to conform to the social terrain of the latest strike. The technology becomes ubiquitous, as transnational corporations and undertakings must be. This flexibility is also rooted in the nature of Pennsylvania's oil boom. Initially, excessive supplies shaped the wasteful practices of the Pennsylvania boom, but in the late 1860s, the defining influence became the growing realization that the industry could never aspire beyond its temporal nature in any single locale. Speculators, drillers, and industrialists all quickly realized that the supply would not last. This became the primary influence in defining land-use practices.

Today, if one could see beyond the steep walls of this valley, one would see that the land around Oil City remains a hub for the storage and refining of heavy crude—but barely. Huge circular tanks crowd the slight road that winds through the narrow gap affording the only passage north. The tanks bear the emblems Quaker State or Pennzoil and hold only the heavy-grade crude now brought from the much-used wells. Where there was once unlimited opportunity for employment and personal financial speculation, there are now few industrial jobs and many fears about the remaining industry's exodus to the American South.

Today, the industry here wrings out the earth to try to extract a bit more crude. Where once the oil gushed from the ground, natural gas is now pumped down to flush out the remaining oil in each geological pocket. These problems, and locating a safe supply of drinking water, now occupy the technology and resourcefulness of residents. The boomtowns that filled this narrow valley during the 1860s are gone, and the young forest of the Oil Creek State Park seems to wrap itself around the remnants of the region's history in order to conceal them for posterity.

Who could have imagined that this valley would be left as an artifact of a bygone industry just over one century after the occurrence of one of the most incredible industrial booms ever seen? During the 1860s, the Oil Creek valley was the world's largest producer of oil. Towns such as Franklin, Titusville, and Oil City orchestrated control of the world's supply of oil, when they had been sleepy hamlets only months prior. This valley was the site of the *first* oil boom in human history. For a short time, this valley was oil, and oil was this valley. The story of this episode grows out of human resourcefulness and courage but also from a culture of massive development and exploitation. The meaning of the Oil Creek valley rings as a cacophony of human and natural forms, past and present, which have worked in harmony and opposition throughout the history of this place. This book tells the story of one period in this place's history, resulting in the discovery and development of one of the world's most important commodities. Yet, like all such locales, the character of the place extends well beyond its commercial value.

Chapter One

"A Good Time Coming for Whales"

The bearded man often wore a black top hat, like the era's other men of
stature. His stature, however, was a fleeting deception. Edwin Drake would
always be in and out of financial difficulty. His work for the railroad paid well
enough, but the death of his wife brought new responsibilities. His own
health worsened with stress and age, and finally the doctor urged him to
move to a more rural area. That prescription drove him to accept an offer to
leave his conductor position in Connecticut and travel to western Pennsylvania in 1858. His major qualification for the new job: a pass for free travel
on the rail lines.

Drake was neither a man to take risks nor a poised man of success. His
new employers packaged him like a presidential candidate, even reshaping
his appearance. To his head and body, they added the hat and suit. To his

name, they added the label "Colonel." Such details were necessary, the Seneca Oil Company believed, if Drake were to get the assistance of local residents near Titusville, Pennsylvania. He would need help to gather materials as well as to hire an assistant who would actually drill the first well for crude oil. In reality, Drake needed much more than such window dressing to win the public's support.

In the coming years, Drake became at once the early industry's preeminent hero and goat. His persistence resulted in the first oil strike; his lack of interest in the business then left him destitute as others applied his ideas to make millions. Maybe someone else would have struck a well if Drake had not. A progression of development had begun, in which Drake was simply one cog. But the facts bear out that Drake's efforts initiated a period of monumental change in the Oil Creek valley and in industry generally.

Historians and oil buffs place the striking of his well, which came in on August 27, 1859, with that of the 1901 Spindletop strike as watersheds of the industry. While the event certainly warrants such distinction, it overlooks the prehistory and context of Drake's strike. His efforts grew out of the technological evolution of products used for lubrication and illumination. Without the existing markets and processes for such products, Drake would have never been sent to Pennsylvania's hinterlands. These developments combined with a growing recognition of western Pennsylvania's odd blessing to attract the attention of industrial development.

More than any other product, whale oil whetted the human appetite for clean, efficient, and affordable illumination. Operating out of the northeastern United States, and later Hawaii and the northwestern United States, the American whaling fleet defined the process of the hunt and refining technology, thereby establishing dominance over the fishery. Most importantly, this dominance established trade markets to disperse whale oil illumination internationally. Illumination proved to be so integral and basic a technology that Americans readily purchased the expensive oil. This success, however, aroused a grassroots desire for a less expensive alternative.[1]

The relationship between sperm and petroleum oils suggests a great deal about differences in industrial development and technological practices before and after the Civil War. While the product of each technology is similar, the processes could not be more different. With the start of massive petroleum exploration and other similar energy developments, 1859 marked a good time coming for the kindly sea creatures hunted nearly to extinction.

Like noiseless nautilus shells, their light prows sped through the sea; but only slowly they neared the foe. As they neared him, the ocean grew still more smooth; seemed drawing a carpet over its waves; seemed a noon-meadow, so serenely it spread. At length the breathless hunter came so nigh his seemingly unsuspecting prey, that his entire dazzling hump was distinctly visible, sliding along the sea as if an isolated thing, and continually set in a revolving ring of finest, fleecy, greenish foam. He saw the vast, involved wrinkles of the slightly projecting head beyond. Before it, far out on the soft Turkish-rugged waters, went the glistening white shadow from his broad, milky forehead, a musical rippling playfully accompanying the shade; and behind, the blue waters interchangeably flowed over into the moving valley of his steady wake; and on either hand bright bubbles arose and danced by his side.[2]

There are many moments of note during the typical whale hunt, yet none more suggestive than "the dive." The hunter-prey relationship of this harvest is particularly fascinating because of the differing sizes of each entity, but additionally because the hunt marks the convergence of these two beings' opposing habitats, land and sea. The dive constitutes the moment of contest to see which realm will emerge victorious. The whale dives after the harpooner has struck and set a harpoon in the whale's gelatinous flesh. The reaction of the prey is to flee and hide within its sanctuary, the deep sea. To defend against this reaction, the hunter attaches a line to the harpoon that will allow the retreat to take place—but only to a point. Then the diver becomes transporter of a dinghy holding six or seven adult men. This trek is called "the Nantucket sleigh ride."

The dive must inevitably end, and the men are readied with additional harpoons when the prey surfaces for breath. The whale hunt is a crude process; yet its primitive technology creates a certain attraction. The contest appears as a fairly equal one, in which the foolhardy sailors risk life as does the behemoth mammal. Such an interaction gives the scene a romantic beauty. In the quotation above, Herman Melville never mentions that the scene is a hunt nor that a whale has any independent existence at all. Human ingenuity and tact simply strain against the wit and power of the prey in a beautiful, symbiotic interaction.

The passage describes the great white whale that has lured the crew of the Pequod far from the waters of New England. The beauty of the powerful forces of nature even overshadows the violence of the actual hunt. However, among all the layers of delicate description Melville and his readers never lose sight of this relationship between hunter and prey. It is a relationship based in the acquisition of a valuable commodity. Ahab does not need the whale for his own sustenance; he harvests a resource for which Americans are willing to pay. Despite their commercial undertaking, however, once they spot the whale Ahab and his crew become involved in a test of natural strength and resourcefulness against a powerful foe.

The cultural and economic value of illumination spurred the pursuit of whales as well as that of a consistent supply of petroleum. Nineteenth-century Americans made such illuminants critical to the life they wished to lead. Additionally, lubrication became a necessary aspect of evolving systems of industry. These uses made such oils valued commodities and allowed the pursuit of each to become extensive industries in their own right. Though it would lead to more far-reaching changes, the collection of whale oil in America during the late seventeenth, eighteenth, and nineteenth centuries soon defined products and processes very different from any previous sources of energy and illumination.

Before it could influence other industries, the pursuit of whales needed to pass from the status of fishing, as for cod, to industrial production and processing. The American whale fishery lasted from 1712 to 1860 and reached its zenith in 1846, when the fleet numbered 736 sailing ships.[3] The whaling towns supporting the industry operated as unique frontier outposts. While they defined New England's coastal culture for decades to come, they also established the infrastructure for the distribution and use of illuminating oil throughout the nation.

As early as 1688, the industry developed out of the Massachusetts Bay Colony, although there had been scattered efforts at hunting whales throughout the coastal colonies since 1647.[4] Nantucket Island proved to be the most famous of these sites, particularly as the launching point for the near-to-shore hunting that lasted until nearly 1726. During that year, a drop in the whale population drove ships out to deeper water.[5] These men became the "Nantucketers," whom Melville depicts as controlling the seas of the world.

The whalers first sought the right whale, but in 1728 the extending fishery area brought one Nantucketer across the migratory path of a different beast, the sperm whale. This kill shifted the entire fishery's focus toward the sperm, which possessed significantly more oil and an important pocket of oily spermaceti (originally believed to be sperm) in its head, which, when frozen and pressed, proved a remarkable substance from which to make candles. If the industry concerned illumination, the hunt now had to be for the sperm.

Each sperm whale proved to be the equivalent of a small gusher. The whale boat steered the carcass to the mother vessel, normally called the "cook boat," where the crew lashed the beast to the vessel's side and began "cutting in," which involved peeling the blubber off in strips.[6] Drained of its pockets of oil, each animal's blubber and meat was then chopped into small cubes and boiled down in stoves on the deck to produce an additional supply. Normally one whale produced five hundred gallons of oil or more. By the early 1800s, whaling vessels collected thousands of gallons of oil during three- to five-year voyages. While the takes on such voyages varied greatly, sixty to seventy kills constituted a satisfying voyage. By 1850, roughly $3 million of capital supported this industry, which created a product valued at nearly $8 million annually.[7]

During the 1800s, the whale population steadily decreased as the efficiency of the hunting process grew. The decline forced hunters to stray farther from their home waters. By midcentury, the fishery focused almost exclusively on the Pacific and Arctic waters.[8] The population figure, however, was only one of the variables contributing to the decline of whale oil. For instance, historian Walter Tower downplayed the importance of the mammal's waning population. Instead, he considered the discovery of an alternative fuel the single greatest contributor to the industry's demise. "Even in the face of the other unfavorable conditions," he wrote, "the fishery must certainly have prospered but for the discovery of petroleum. The population of the country was increasing; the people would have had light without much regard to the necessarily high prices of oil, and the market demand would undoubtedly have increased beyond the supply."[9]

At this critical moment, Drake captured flowing petroleum in the Oil Creek valley. Although investment from whaling supported a great deal of the initial development, there was no denying that petroleum would signal the end of the expansive search for whale oil. According to Tower, "The struggle for supremacy was fierce but short and ended in the only way that it could—in favor of the better, more easily obtained and then seemingly inexhaustible kerosene."[10] In 1861, *Vanity Fair* published a cartoon depicting formally festooned whales attending a ball honoring Drake's well. As they danced and frolicked in celebration of the technological progress that had freed them, the whales knew the time had passed when the world's need for illumination lay on their flukes. A good time had come at last.

While lacking the attraction and excitement inherent in the whale hunt, the discovery of flowing supplies of oil in the Oil Creek valley dazzled many on-lookers. Petroleum occurred as an inanimate mineral in pockets beneath the earth's surface, not as a breathing creature seeking to elude or destroy the human attempting its harvest. Interestingly, however, some of the early wit-nesses tried to impose similar attributes upon the earth's geological struc-ture. Often, the earth appeared alive in its efforts to resist the speculators' ef-forts.

The process of mining oil from the earth proved unnerving to more than a few residents, particularly due to the odd methods of extraction. Observers wrote that "the earth seems to bleed like a mad ox, wrathfully and violently." Jim Burchfield, editor of the *Titusville Gazette*, upon observing the striking of one of the first wells in the region, wrote this account:

> We have no language at our command by which to convey to the minds of our readers any adequate idea of the agitated state at the time we saw [the well]. The gas from below was forcing up immense quantities of oil in a fearful man-ner and attended with noise that was terrifying. . . . When the gas subsided for a few seconds, the oil rushed back down the pipe with a hollow, gurgling sound, so much resembling the struggle and suffocating breathings of a dy-ing man, as to make one feel as though the earth were a huge giant seized with the pains of death and in its spasmodic efforts to retain a hold on life was throwing all nature into convulsions.[11]

In this passage, the imposition of human attributes upon nature illustrates the discomfort with which some reacted to these new, increasingly intensive industrial uses of the surrounding landscape. However, it also suggests a new paradigm for illumination technology.

In Pennsylvania, a more businesslike exchange replaced the volatile hunter-prey relationship of the whale hunt. Perfecting the technologies for gathering, storing, transporting, and refining the crude represented the challenge to the new petroleum industry. As an extensive interaction be-tween human technology and the natural world, the petroleum business cre-ated industrial repercussions much broader than had the whale hunt.

Most importantly, entirely new groups orchestrated this confrontation: speculators and marketers well-schooled in the nation's energy and illumi-nation needs drove the search for a steady supply of petroleum. Pennsylva-nia rock oil was a commodity not limited by the amount of fats and oils that could be extracted from a single mammal, however large it might be. Instead,

the greatest mystery in rock oil's discovery lay in the amount available beneath the earth's surface. It truly seemed unbounded—if it could first be
brought to the surface.

The process of taking a resource that occurs in abundance and finding a
use for it can take many years—sometimes thousands of years, as in the case
of petroleum. This process depends on a resource's similarity to existing
products already in use and on the universality of the need that the resource
fills. The seizure of whales, for instance, increased as Americans learned the
value of whale meat, oil, spermaceti, and finally bones. Through the use of
sperm or whale oil, Americans learned the commercial value of making easy
illumination universally available. Their acceptance of this cultural change
cleared the path for the introduction of a variety of other energy sources.

Such an episode of establishing a resource's market value is part of a
process called *commodification*.[12] In such a process, resources attain value
based on the existing culture's fluid and changing need or desire to possess
them. The value of an object or a substance is fully governed by the surrounding human culture. This becomes particularly acute in a market-based
society, which places fluctuating prices on resources. Within such a market,
few technologies had the universal appeal of illumination. And sperm and
other oils had prepared the economic path wonderfully.

In the example of the Oil Creek valley, the crude had been an obvious part
of northwestern Pennsylvania's lore and geology since the area's earliest settlement by humans. From being a nuisance to becoming a treasure, the ooz-
ing crude acquired a price and market for the occupants of this region; indeed, it became a trademark of the place. These shifting definitions reveal a
great deal about the culture in which they were created. They also indicate
petroleum's dependence on similar products being used during the eighteenth and early nineteenth centuries.

The products used for illumination shared many qualities, but primarily
all required burning. Animal and vegetable fats were the only known sources
of burning oil in the early 1700s, making illumination a messy affair of
smoke, uneven flame and light, unpleasant odors, and the incessant chore of
wick trimming. None of the oils were pure enough to burn very well, but
whale oil was the most dependable. Additionally, unlike other fuels, whale
oil burned in a variety of lamps.

From 1830 to 1850, rapid industrial growth fed the furious development
of illuminants. These changes included a near doubling of the population,

increased urbanization, a massively expanding factory system, and the addition of some nine thousand miles of railroad. Progress toward meeting these increasing needs occurred on a variety of fronts: distillation of illuminating oil from turpentine; expansion and technological changes in the manufacture of lard oil; growth of the domestic gas industry for illumination; and pioneering work in the production of coal oils, particularly in England.[13] Each of these markets found persistent niches, but petroleum's viability as an illuminant solidified as it replaced these more expensive sources.

The first breakthrough in this development involved the patent for camphene taken out by Isaiah Jennings in 1830. The idea of redistilling spirits of turpentine, which would burn alone or with alcohol, was the only major illumination advance to take place in the United States before 1850.[14] Camphene's affordability made it the prevalent illuminant in the United States during the later 1840s, when shortages of whale products made them too costly to remain competitive. While there were still whales in the ocean, the fishery had been forced to move to the Pacific or to other very deep waters, thereby increasing production costs and the final price of the oil.

In addition to camphene, a new lard oil developed in the pork industry of Cincinnati and made the solar lamp the outstanding table lamp in antebellum America. This lamp's design would become standard for use with all oils. Interestingly, its greatest attribute was its ability to burn a variety of oils—thereby allowing the user to choose the more affordable alternative. This lamp allowed lard oil to seriously challenge sperm oil's dominance in American homes during the early 1840s. It also freed up the illumination market for newly developed fuels.

Such competition made some businessmen consider the well-known rock oil, petroleum. In 1847 James Young of Glasgow, Scotland, began illumination experiments with petroleum from a water spring in Derbyshire. The supply soon ran out, and Young shifted to the sale of illuminating oil made from the distillation of coal. Young evidently never considered methods to seize more crude from beneath Earth's crust.

Coal oil proved to be a competitive illuminant, but the lack of an adequate lamp limited its use until 1856. In Canada, Abraham Gesner obtained oil from bituminous coal in 1846 and called it "kerosene." This same name quickly extended to all illuminating oils made from minerals.[15] Coal oil became popular during the 1850s. Although its foul odor kept development relatively slow, by 1859 nearly sixty firms were manufacturing coal oil at an investment of nearly $4 million.[16] So, whereas Tower gives sole credit to petroleum for defeating the prominence of sperm oil, the assault was in fact a piecemeal attack by many products.

One other illuminator received attention during this period, but its form differed from the fluids listed thus far. Combustible gas produced by heating coal had been known since the seventeenth century, with the first commercial manufacture occurring in 1802. Gas produced a much superior light to oils, which enticed Americans to put it to experimental use. In 1816 Baltimore, Maryland, installed gas lines in one section of the city and began the first large-scale American use of minerals for illumination. By 1830 Boston and New York had adopted Baltimore's model. By 1850, fifty-six plants operated both in urban centers and in factories in outlying districts.[17] While gas made from coal grew faster than any other illuminant of this era, it required a great deal of infrastructural investment. In addition, company owners made the resource inaccessible to many Americans by keeping the price high even after start-up costs had been recovered. Therefore, gas use also spurred the nation's search for a cheaper alternative.

Each illuminant helped bring light to darkness. However, each product left dramatic room for improvement. While each development functioned to lay the groundwork for the rapid acceptance of petroleum upon its "discovery," the coal oil industry, which grew significantly in the United States during the 1850s, achieved a national distribution network that could be shifted most easily to other fluid energy commodities. From 1857 to 1859, experiments verified that petroleum could be used instead of coal in the same distillation process and distributed over the same network of users. These findings, however, were filed away in case a sustained supply of petroleum oil was ever secured.

The rapid commercialization of petroleum, or rock oil, constitutes a striking example of cultural commodification. Its occurrence and location had been known for many years prior to Drake's successful drilling, yet its potential had gone unexplored. Settlers initially named the oil occurring along Oil Creek for the Seneca people, who were the native inhabitants of this portion of North America at the time of European settlement. The Seneca were thought to have been the original human inhabitants of this place, so its product should therefore bear their name. This region, however, had been at least a temporary home to mobile peoples centuries prior to the Seneca.

While these earlier peoples left few written records, historians position them inhabiting North America during the Hopewellian period, which lasted from 200 B.C. until A.D. 800. Archaelogists have found village sites connecting these riverine travelers' homes along the Mississippi and Tennessee Rivers to more temporary sites in present-day New York.[18] These early agriculturalists ventured from their settlement sites in order to collect necessary resources. A small stream above Pittsburgh, just off the Allegheny, became such a frequent stop as word spread of its freely pooling oil. This oil could be used for decoration, skin coloring, and other ceremonial rites.[19]

While mobility distinguishes these early Americans from later peoples, they were also industrious and highly regimented. Initial European explorers in the valley found long, narrow troughs that had been dug along Oil Creek just below its junction with Pine Creek. Roughly two thousand troughs were found scattered over this level plain, and others could be found at intervals throughout the Oil Creek valley. Each one spanned seven or eight feet in width and six to ten feet in depth, shaped as circles, oblongs, ovals, and squares. These troughs were cribbed with lumber, which had been preserved by the oil stored in them. Large trees growing out of the troughs suggested to early residents that the troughs had lain dormant for many years, possibly even centuries.[20] No evidence remained in the vicinity of Titusville as late as 1847 to show from where the timbers for such cribbing might have come.

As agriculture became more regular, after 1000, Iroquoian people settled in this area of northwestern Pennsylvania. During this period, the Seneca established their own proto-Iroquoian language.[21] They volunteered no knowledge of the pits and did not use such technology in their own collection of oil. In the absence of any direct testimony, the Reverend Eaton's 1866 history analyzes the construction of the pits and suggests that they were built to allow the oil to seep from the saturated soil and into the troughs for collection. This process supposes that the pits were left unattended for long periods of time while they filled. The oil would then be collected on excursions from the users' home regions.

Eaton suggests that such methods contradicted the Senecan view of natural order; however, this is not to suggest that the Seneca neglected to value the abundant resource near their homeland. They were indeed the first people of record to place a value on the thick, black film filling nearby pools and often spreading congealed clouds over the creek's surface. The Seneca skimmed the oil from the water's surface, using a blanket as a sponge or dipping a container into the water. Once collected, the brownish crude served as an ointment or a skin coloring, but nearly always was used only externally. Many early European explorers noted the important role the substance played in Senecan culture. Such explorers also called this odd stream Oil Creek.

From its earliest notice—whether by aboriginal or by immigrant peoples—the geological substance so near the earth's surface identified this locale. Oil Creek and the fluid pooling around it were first recorded as a detail of Lewis Evans's *Map of the Middle British Colonies in America* in 1755 (map 1.1). Additionally, very close to the present site of Titusville and Oil City, the word *Petroleum* is printed on Peter Kalm's *Map of New England and the Middle Colonies* in 1772 and Thomas Pownall's *Map of the British Colonies in North America* in 1776. A Moravian missionary recorded the first known written observation of oil in this region in 1768.[22] In this account, the missionary observed a number of different types of oil springs. He also noted that the Indians preferred those springs feeding directly into the creek, and he wrote of their method of dipping the crude from the pits and then boiling out the remaining water.

Passing American soldiers of the Revolutionary era often noted the remote area. In 1775 General William Irvine made a trek to the region specifically to explore along Oil Creek and filed the following report: "It has hitherto been taken for granted that the water of the Creek was impregnated with [the oil], as it was found in so many places, but I have found this to be an error, as I examined it carefully and found it issueing out of two places only . . . on opposite sides of the Creek. It rises in the bed of the Creek at very low water, in a dry season I am told it is found without any mixture of water, and is pure oil."[23] By modern thinking, a stream of pure oil should have begun a great land rush to the Oil Creek valley. However, for a 1780 observer, this fact merely classified the region as poor for agriculture and an unpleasant place to reside.

Such reports continued with General Benjamin Lincoln's observation of a 1783 incident when his soldiers stopped at the springs, "collected the oil, and bathed their joints with it." This, he continues, "gave them great relief, and freed them immediately from the rheumatic complaints with which many of them were affected." The troops drank freely of the water, which, by and by, "operated as a gentle purge."[24] Thought of such a "gentle" purge makes most modern observers cringe and wonder whether the general had

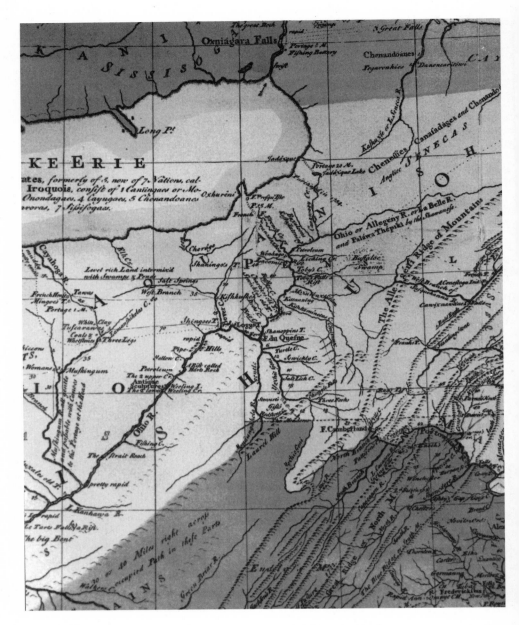

Map 1.1. Lewis Evans's *Map of the Middle British Colonies in America*, 1755

participated in or just observed the imbibing. Either way, the region's oil enjoyed a growing reputation for its medicinal capabilities. Accounts of this natural curiosity, based upon Lincoln's observations, were included in Jedediah Morse's *American Universal Geography* in 1789 under the heading "American Natural Curiosities," in *Massachusetts Magazine* in 1791, and in Joseph Scott's *United States Gazetteer* in 1795.

Settlers in the region soon began to gather oil from springs on their property by constructing dams of loose stones above the water's surface, ten to fifteen feet in diameter, around the place where the oil bubbled. Dams created an eddy inside the wall that confined the floating oil, while the water flowed out freely between loose stones.[25] The oil accumulated for several days before being soaked up with a woolen cloth. Ten to twelve barrels of oil might be collected in a season, which was not enough to inspire one to turn the undertaking into an industry. The Hamilton McClintock farm housed the most successful spring, harvesting twenty to thirty barrels.[26]

While not enough to start a boom, such a supply brought petroleum to the consumer for the first time. Nathaniel Carey, one of the first settlers along Oil Creek, brought barrels of oil to Pittsburgh in 1790.[27] During the 1790s, bottles containing the substance also found their way into other urban areas, where they were called "Seneca Oil" and offered as a miraculous cure for many ailments, particularly rheumatism.[28] The American consumer normally ingested Seneca Oil much as he or she would castor oil. Locals and those traveling through the Venango County region were even known to sit in the pools of oil to soak their weary and aching joints. One other potential use had been explored early by the imaginative mind of Benjamin Franklin. Using whale oil and later petroleum in 1857, Franklin began experiments to calm rough waters by dumping the thick substance directly onto the ocean in harbors.[29]

Carey's efforts at distributing the oil were soon surpassed by a young canal boat operator who wished to commercialize the process. In the mid-1840s, Samuel Kier noticed the similarity between the oil prescribed for his ill wife and the annoying substance invading the salt wells on his family's property outside Pittsburgh. This entrepreneur immediately began collecting the waste substance and opened a bottling and merchandising house in Pittsburgh in 1849. The mysterious cure-all, "Kier's Rock Oil," soon sold throughout the northeastern United States. Although he acquired the oil only by skimming, Kier's supply quickly exceeded demand because of the constant flow from the salt wells. With the excess, he began the first experiments with using the substance as an illuminant.[30]

Kier continued to make his former nuisance pay dividends throughout the 1850s. By 1850, Kier was selling an illuminant called "carbon oil" for $1.50 a gallon from a warehouse in Pittsburgh. Afraid of explosion and fire, residents living near Kier's refinery registered complaints with the authorities, who ordered Kier to move his operation from the city.[31] In 1857, A. C. Ferris, a New York businessman, ordered a supply of oil from Kier and began experimenting with its illumination potential. Using his connections from other business as well as aggressive advertising throughout the New York area, Ferris sold around a thousand gallons of illuminating oil in 1858. He cultivated the markets that would allow petroleum by-products swiftly to become the nation's most popular illuminant.

Pennsylvania and other northeastern states entered an age of rapid industrialization after about 1830. Whereas natural resources had been previously connected to subsistence needs, they now came to be viewed entirely separately from cultural ideals such as land tenancy and economic sustainability. Resources no longer would be extracted only for local use. A growing infrastructure for transportation and distribution allowed products to be sold throughout the nation and the world. In Pennsylvania, lumbering and iron production represented some of the earliest industrial undertakings.[32]

Venango County served as a microcosm for many of these developments, possessing an iron furnace and a prosperous lumber and tanning industry until around 1850.[33] The abundance of wood teamed with the available waterways to furnish lumber to much of the state. As this supply diminished near the more populated centers along the eastern seaboard, more and more industries turned to coal power. Coal allowed for more efficient burning, and, unlike wood, was often found near urban areas.

Geologically blessed with the world's largest anthracite (hard) coal reserves, Pennsylvania instituted a massive push toward heavy industry during the rest of the nineteenth century.[34] Preceded by the domestic and industrial use of bituminous (soft) coal, both in Europe and in the American colonies, anthracite's difficulty of extraction and burning slowed its development.[35] Coal would not completely displace fuelwood until 1890, but by 1850 it had become a major source of energy for industry.[36] In addition to mining operations throughout the state, railroads were put in place and, between 1826 and 1842, some 772 miles of canal were constructed. An industrial infrastructure rapidly grew during the nineteenth century, and its locus would be Pennsylvania.

During this era, properly administered technology could make any resource into an economic boom. In some cases, such as the California gold rush of 1849, technology was not even necessary (although it would eventually enhance profits); a substance of known value could be "discovered" by people from a culture that simply knew a better fashion in which to cash in on its value and could transform it into a commodity. The attraction of gold grew more from individual speculation—the personal possibilities of wealth—than from the provision of a critical resource for the growing nation.

This era of expansion required technological innovation to support the new scale of production. For instance, concern over the durability of wooden machinery led to the development of metal gearing mechanisms. In addition, the steam engine as a prime mover extended the use of moving metal parts in factories and other apparatus such as railroads. The supply of one resource became an enticement to related industries to locate nearby. Some industrialists such as Andrew Carnegie began to develop this idea into the vertically integrated company, which would produce supplies of each of its necessary raw materials. The interconnection of industries encouraged the economic primacy of a state such as Pennsylvania, which possessed many of industry's necessary natural resources. The latest resource to be needed was a lubricant that would limit wear on these metal parts and also increase their efficiency. Pennsylvania rock and coal oil became the answer for most industrialists. However, supplies were still limited to surface deposits.

While the powering and lubricating industries created employment and products, no potential use came close to the society-wide benefits offered by illumination. Creating affordable lighting possessed the divine potential of increasing time in the day. By making lighting cleaner and cheaper, more time was literally created in which Americans could develop as a culture and increase their productivity by working longer hours. Thus far, the drive for illuminants had remained separate from industrial developments in energy production. But what would be the outcome if an abundant illuminant could be found and linked with national industrial development? And what if such a resource were also opened up to individual financial speculation? Imagine it: gold's potential for generating individual wealth and coal's importance to the cultural and economic growth of the young nation. The possible wealth to be had would be breathtaking—even potentially unlimited if the infrastructure of the nation were designed with it in mind. As the nation moved toward the Civil War, it was on the brink of discovering such a resource.

Dr. Francis Brewer traveled to Titusville in 1851 in such an atmosphere of economic expansion. As a practicing physician in Vermont, he had two years earlier received a barrel of five gallons of creek oil from his father in Titusville. He became a believer in its medicinal use and often prescribed it to patients. While wandering about his family's lumber holdings in this remote section of the state, Brewer took time to examine a well-known oil spring. Before leaving Titusville for his home, Brewer formally contracted a local man, J. D. Angier, to collect the oil that he noted seeping to the surface. It was the first oil lease ever signed; however, there was still no effort to drill into the earth for the substance.[37] Angier instead dug trenches to convey oil and water to a central basin, where some crude machinery separated the two wells enough to produce three or four gallons of oil a day. Here, then, begins the astounding sequence of chance occurrences that brought Pennsylvania rock oil to market.

Upon his return to Vermont, Brewer took with him a bottle of the oil. The proverbial genie was escaping the bottle. Dixi Crosby, a chemist at Dartmouth College, acquired the sample and shared it with a young businessman, George Bissell, who happened to be visiting his alma mater. Bissell worked in the coal oil business, and the sample immediately struck him as similar. Crosby soon traveled with Brewer to Titusville to examine the spring firsthand. Dr. Brewer later described this moment:

> As we stood on the circle of rough logs, surrounding the spring, and saw the oil bubbling up, and spreading its bright and golden colors over the surface, Crosby at once proposed to purchase the whole [McClintock] farm, which we could have done for $7,000, but [there was] not enough money. When I told Crosby that we [Brewer, Watson and Co.] did not want to take money from the lumber business to put into oil, Crosby said, "damn lumber, I would rather have McClintock's farm than all the timber in Western Pennsylvania."[38]

Upon their return Brewer signed a lease with Bissell to develop the oil occurring on the lumber company's tract of land. The lease, however, was contingent on Bissell's locating financial backing of $250,000.

Bissell began contacting financiers to help support the project. Although each one recognized the oil's potential, they demanded more scientific verification. Even during this age of giddy expansion, the vague promise of money was not enough to attract most speculators. Bissell contracted with Benjamin Silliman Jr., of Yale University, to analyze the substance. In April 1855, Silliman released a report in which he estimated that at least 50 percent of the crude could be distilled into a satisfactory illuminant for use in camphene lamps and 90 percent in the form of distilled products holding commercial promise.[39] Bissell's effort to raise funding suddenly became much easier.

On September 18, 1855, Bissell incorporated the Pennsylvania Rock Oil Company of Connecticut, a corporation founded solely on speculating with the potential value of the oil occurring naturally beneath and around the Oil Creek valley. The value of the valley and its contents began shifting toward an entirely new and unique frame of reference. The process of commodification transformed a nuisance—the oil that seeped into farm land and salt wells—into a product of such value that it would revolutionize life in the Oil Creek valley.

As the story goes, the summer of 1856 brought Bissell a chance look at one of Samuel Kier's handbills, which explained how he brought the oil up from below the earth's surface with salt water (fig. 1.1).[40] Ironically, Kier included such an explanation only to stir enthusiasm among his prospective users; he had no proof of the geological occurrence of petroleum. This mystery had, however, caught the eye of the Pennsylvania Rock Oil Company of Connecticut.

When James M. Townsend, a New Haven banker, succeeded Bissell as president of the corporation in 1857, he contacted Edwin L. Drake, of the New Haven Railroad, about traveling to Titusville and overseeing an effort to drill for crude. One of the few facts of this tale about which there is no disagreement is the lack of any reasonable rationale for Townsend's selection of Drake.

Once Drake had arrived in Titusville, Dr. Brewer took him to the site of the spring, where, Drake would later report, "within ten minutes after my arrival upon the ground . . . I had made up my mind that it [petroleum] could be obtained in large quantitites by Boreing as for Salt Water. I also determined that I should be the one to do it."[41] While it is not verifiable who actually came up with the plan to drill for the oil, the thought occurred to Townsend, Bissell, or Drake at some point during the next year.

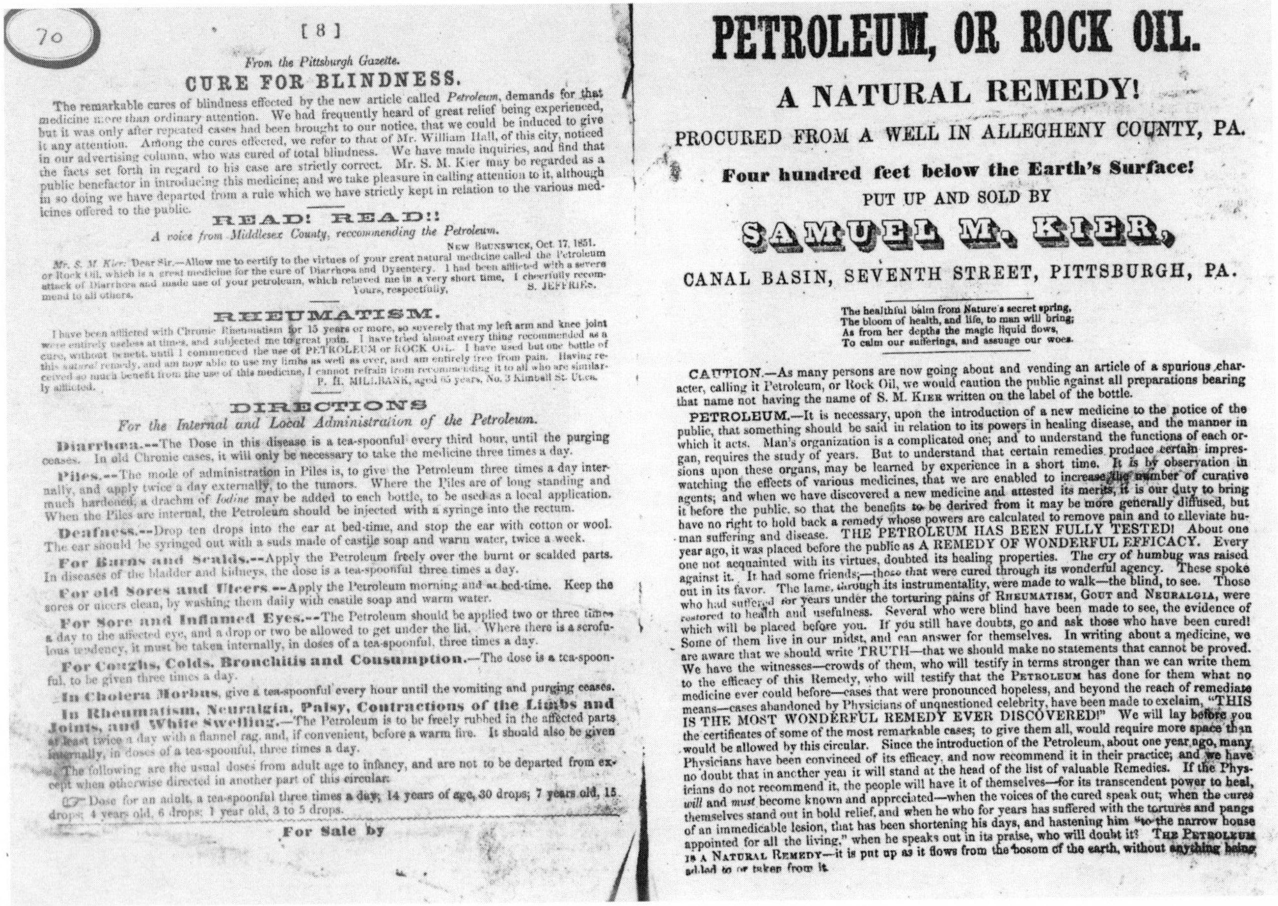

Fig. 1.1. Advertisement for Samuel M. Kier's Rock Oil remedy

On March 23, 1858, Townsend reformed the company as the Seneca Oil Company of Connecticut. Drake brought his family to Titusville in May and began a string of unsuccessful and costly attempts to secure a blacksmith who would drill the well. Drake spent weeks combing Titusville and then the surrounding towns in search of help with his project. Men were not willing "to work for a lunatic," he remembers in an 1879 account. Refused at every turn, Drake finally traveled a hundred miles away and found a salt borer with whom he entered into a contract for a thousand feet of boring. Drake then returned to Titusville and waited for his employee—but to no avail. It turns out that the man thought the easiest way to get rid of Drake was to make a contract and pretend to come. Another hired hand died en route. As Drake faced the approaching winter in October, he decided to wait until spring to continue the experiment.

When spring arrived, it brought Drake his first hopeful return. A contact he had made in Tarentum, over a hundred miles away, sent one of his own employees to assist with the drilling. William A. Smith, a trained blacksmith who had also been involved in boring for salt, arrived in April 1859. By Smith's account, Drake asked him what he thought the operation's chances were. "Very good," Smith reports that he replied. "I would not be afraid to guarantee . . . ten barrels a day. [Drake] said, stop! half that will satisfy me."[42]

Smith's account of assuring Drake that drilling would be successful remains extremely dubious, considering the widespread regional perception of Drake as a lunatic. Drake's own story is also difficult to trust or verify. However, it is obvious from every account that Drake repeatedly displayed a degree of patience beyond the normal. Again and again, the piping broke off and Drake drove forty miles by wagon to Erie for more. And then there were the relentless local hecklers. Smith drilled the first well for oil amid frequent catcalls from townspeople who had ventured out from Titusville.

Even before drilling, Drake and Smith immediately began cribbing the sixteen-foot hole that had been dug over the oil spring. After experiencing frustration with water entering the hole from the creek, one of the two men decided not to crib the hole any more but only to drive an iron pipe the rest of the way and begin drilling.[43] Before long the novelty of the project had worn off for Drake and Smith, for the townspeople, and, worse yet, for the investors in New Haven. During the late summer of 1859, Drake ran out of funds and wired to New Haven for more money. They offered only funds for a trip home—the Seneca Oil Company was done supporting him in this folly.[44] Drake, showing remarkable courage, took out a line of credit locally and decided to stay on for a bit longer. He wrote Townsend, "You all feel different from what I do. You all have your legitimate business which has not been interrupted by the operation, which I staked everything I had upon the project and now find myself out of business and out of money."[45] Drake's heartrending note reflects his desperation to succeed in this endeavor. It would be a similar sentiment that brought the string of boomers following his success. On August 27, the drill dropped five feet after one kick, evidently breaking through some sort of underground crust (fig. 1.2).

As is the case with many moments of historical importance, when the well came in—which is a misnomer for a seeping well such as this, not a flowing one—there was actually no one in attendance. Ever pious, Drake stopped drilling on August 27 to observe the Sabbath. He returned to Titusville while Smith remained. The engine house had been split by a wall and the Smith family made the single room its home, so Uncle Billy, as he was called, never truly took a day off from the site.

On Sunday, August 28, 1859, Uncle Billy and his son, Samuel, checked the well site for any changes. A dark green substance with a yellow tinge floated freely around the wellhead. Uncle Billy crouched there with Samuel and lowered a piece of tin rain spouting down the narrow pipe. The future of the valley was held in this improvised ladle as it wobbled through the piping back to the surface. Uncle Billy peered into the tin ladle and saw only a dirty, greenish grease.[46] He had no doubt as to what he had just discovered: for the first time in human history, oil had been intentionally struck beneath the earth's surface. Drake was called from town; when he arrived, Smith called to him, "Look at this!" Drake approached and then asked, "What is that?" Smith responded, "That, Mr. Drake, is your fortune!"[47]

The *New York Tribune* carried the earliest accounts of the well on September 13 (believed written by Brewer).

Last week, at a depth of 71 feet, [E. L. Drake] struck a fissure in the rock through which he was boring, when, to his surprise and the joy of everyone, he found he had tapped a vein of water and oil, yielding 400 gallons of oil every 24 hours. The pumps now in use throw only five gallons per minute of water and oil into a large vat, when the oil rises to the top and the water runs out from the bottom. In a few days they will have a pump of three times the capacity of the one now in use, and then from 1,000 to 1,200 gallons of oil will be the daily yield. . . . The excitement attendant on the discovery of this vast

Fig. 1.2. Edwin Drake, *right*, and Peter Wilson, Titusville druggist, in front of the first well, 1860.

source of oil was fully equal to what I ever saw in California when a large lump of gold was accidentally turned out.[48]

Unknowns dominated the scene of oil: How much occurred beneath the earth's surface? Was it worth anything? What could be done with it? How would the discovery change regional life? However, amid such uncertainty, one fact predominated: Drake, who had been widely ridiculed for his attempt to drill for oil—commonly referred to as "Drake's folly"—was vindicated. A steady stream of onlookers rode to the site from throughout the valley and the broader region. They came to gawk in wonderment but also to calculate how this discovery might be relevant to their own futures. From the outset, the speculation centered on individual enterprise—much like the gold fields so prevalent in the national memory.

At this time oil had little if any commercial value. As one correspondent described him, Drake was "then in the position of the man who drew the elephant at the raffle, and did not know what to do with it after he got it."[49] From this point, the tale veers significantly from its human conveyors. It remains easy to miss this trail and to study the history of petroleum only as a human accomplishment. Yet the truly historical importance of this discovery had little if anything to do with the personalities who first developed the method for oil's extraction. The historical significance here, the episode that made this scene one of revolution, came with the ensuing change wrought upon the landscape of the Oil Creek valley.

Writing in *The Republic of Technology*, Daniel Boorstin stressed that technological developments have rippling effects on other details of life, leaving little unchanged: "each grand change brings into being a whole new world. But we cannot forecast what will be the rules of any particular new world until after that new world has been discovered."[50] The discovery of methods of bringing oil from the earth in great amounts made the Oil Creek valley one of these "new worlds" of which Boorstin wrote. The roots of the Oil Creek valley's development, however, reach past technology to the cultural values of residents fueling development.

The oil that had once acted as a mere identifier for a region dependent on agriculture, iron, and lumber suddenly became the vehicle for the region's rise to prominence and economic progress. What had once been valueless now was a commodity of skyrocketing worth—locally, regionally, and nationally. In this valley, invisible and beneath the surface of the earth, lay a tremendous natural—and indeed national—treasure. In the early years of this industry the reaction of occupants and onlookers indicated the priorities that would govern the use of the valley's resources over the next decade.

The nation and the industry made the first focus of their agenda to find uses for the enormous amounts of crude. A letter written to *The Living Age* by a recent visitor to Petrolia leaves no doubt that valuable uses would soon emerge. "Such is the value of the oil," he wrote "that from the commencement the demand had been [in] advance of the supply, and it is sought with avidity by men ready to pay cash at the wells for every gallon." The editors of the magazine extended the writer's excitement by responding that the wells were "certainly one of the wonders of the world" and that such encouragement predicted "a good time coming for whales."[51]

In reality, the good time for whales was a ways off. Scientists struggled throughout the decade to perfect uses for petroleum, even the proper distillation to allow for its clean burning as an illuminant. Much like magicians, chemists stationed in their laboratories varied methods of distillation and refinement in an effort to create an assortment of by-products. Most important of these new products was gasoline, which entered commercial usage around 1863.[52] Largely useless as an illuminant, gasoline powered the airgas machines that became popular in factories and mills by the early 1870s.[53]

The most persistent effort of this era, however, focused on making petroleum an energy source to compete with wood and coal. Fuel oil possessed the most potential for fulfilling universal needs, including use in experiments in ships and locomotives. Lastly, petroleum fed one nonburning use, the lubrication of metal mechanical parts. While petroleum became an acceptable alternative for each of these uses, as well as others, the availability and dependability of the supply needed to be assured. No doubt, though, petroleum's moment had arrived.

The nation's best-known science and business journals began actively considering petroleum's future in the early 1860s. In 1862, both *Merchants' Magazine* and the *Journal of the Franklin Institute* (*JFI*) began a decade-long search for uses of the abundant petroleum. Both journals focused on experiments to perfect methods of petroleum distillation that would leave a grade of fuel that would serve as an effective illuminator. Economics motivated *Merchants'* interest in finding such a method. While this was also an impetus behind the *JFI* coverage, this journal also possessed a genuine scientific agenda.[54]

Throughout the mid-1860s, the journals published findings of scientists experimenting with methods to better refine the oil for use as an illuminant. *JFI* added a new twist to the topic in 1872 by addressing consumers' concerns over product safety. However, first the writer saluted the oil industry and the "epoch-making period . . . [of] the material progress of our race that it has ushered in." Aside from all its other uses, the author stressed that widespread illumination had changed our world by making possible an "increased culture."[55]

The good time had come for whales, but a very different time had also taken shape. The technology of industries such as the collection, refinement, and distribution of whale oil seems truly of a different age than the scale, scope, and economics of the world of oil. From being a nuisance, oil became one of the world's most integral and valuable commodities in little more than a decade. Drake, the bearded man who reportedly did not even recognize crude when he saw it, had applied folk knowledge to the industrial age. The era's businessmen and popular culture then took his discovery and created a commodity from a resource. These creations then led to others until a new world grew from this greasy place.

For centuries, the place had been known as a geological anomaly, but it would require an ecological revolution for a human generation to realize the resource's potential. Their ideas about how to use this resource helped to define a new age for industry. The repercussions of this revolution ripple to the present, when petroleum's periodic shortages so alter everyday life that entire cultural patterns change, and when the potential of an unfriendly nation controlling its supply could drive the United States to war.[56] This, as each detail in this book, composes a portion of the process of commodification. In the case of oil, the process had begun in a piece of bent tin.

The whole earth is the Lord's garden, and
he has given it to the sons of man upon a
condition (Genesis 1:28): Increase and mul-
tiply, replenish the earth and subdue it. . . .
Why, then, should we stay here striving for
places to live . . . , and meanwhile allow a
whole continent . . . to lie empty and unim-
proved?

—John Winthrop, *Old South Leaflets*

It is certain . . . the development could
never have gone on at anything like the
speed that it did except under the American
system of free opportunity. Men did not
wait to ask if they might go into the Oil
Region: they went. . . . It was a triumph of
individualism. Its evils were the evils that
come from giving men of all grades of
character freedom of action.

—Ida Tarbell, introduction to Giddens,
Birth of the Oil Industry

Chapter Two

"A Triumph of Individualism"

Slack-jawed townspeople in Titusville greeted the announcement of Drake's discovery with disbelief. Oil brought up from the ground? His idea for drilling had really worked? It was an affront to everyone who had lived here for decades without making such an attempt. But now that the fantastic moment had occurred, this "discovery" could also serve as an opportunity for each. One by one they stole out to join the parade of onlookers streaming by the site of Drake's well. They would first go see it for themselves and then determine how to proceed.

For one figure, however, this procession served as no less a business opportunity than Drake's well. This figure stealthily moved in the opposite direction from the group. He mounted the fastest horse that he could locate and headed south along Oil Creek.[1] He would ride the beast until it foamed with sweat and exhaustion. In a time of shifting priorities, many nineteenth-century onlookers might, in fact, compare this rider's imperative with that which had driven another rapidly moving mount through the countryside. Whereas Paul Revere had ridden through the Massachusetts colony to inform inhabitants of an approaching military force, this figure bore the tidings of a different kind of invasion. In fact, the ink pen clutched in one hand placed the rider within the invading force. This Paul Revere represented the first wave of the invading capitalists that would overrun the Oil Creek valley during the next decade.

Jonathan Watson, the pen-wielding rider, represented the lumbering firm of Brewer, Watson Company, which had leased Drake his famous plot of land. More importantly at this moment, he possessed two unique abilities: he could ride a horse very well and could easily strike agreements with the valley's occupants who were familiar with him. While the rest of Titusville stood by Drake's well and pondered the future of the valley, Watson directly seized it. Watson immediately hypothesized that the oil must be most readily available from the lowland areas along the stream. During the next few days, he visited each of the forty-three German and Scotch-Irish farmers whose land bordered the creek along this valley.

A businessman in the area since 1845, Watson knew the physical geography of the valley as well as the personal geography of its residents. He spoke with the landowners and had little difficulty in obtaining leases at bargain rates. If the people knew anything of Drake's undertaking, they had no idea of their own land's skyrocketing value. This man at their door offered them money not to buy their land but just to *use* a portion of it—a portion most often along the river and of little agricultural value. For most, it was an easy decision. Indeed, they may have thought they were swindling *him*. By the middle of September, he had leased much of the most promising land along Oil Creek. Watson would become the region's premier oil producer, drilling more than two thousand wells by 1871.[2]

In reality, Jonathan Watson simply practiced "good business" on this ride in which he chose not to tell residents how the value of their lands had suddenly changed. American capitalism rewards such resourcefulness. During the 1860s oil boom, such entrepreneurship defined the character of this place and its industry. Watson and others like him guided development of both the industry and the region with any rationale that they chose. Most often they chose the most expeditious route toward personal fortune. In this tendency, the oil industry is not unique; however, the occurrence of oil proved uniquely intractable.

The strictures or regulations emanating from the priorities of participants defined the oil industry, as they do other undertakings as well. In the case of oil, the cultural and social forms took place around the lack of interest in limiting development. Most important, the legal system offered little help in controlling it; in fact, the system of land law only furthered a laissez-faire approach to development and land use. Transient labor, long-distance financial speculation, subleasing, land abandonment, and overdrilling began as details of the early industry's temporary carelessness in this, its boom period; eventually, however, such details were institutionalized within the rule of capture, the only law guiding oil speculation.

First applied on this continent during the colonial period, the American property system, while effective in controlling settlement and agricultural development, failed miserably in organizing the extraction of oil. Based on John Locke's directive that land ownership should depend on individual enterprise and labor, the American system was also influenced by Adam Smith's call for trust in the natural liberty of individuals to own and use land as they wish.[3] Locke wrote that in the original state of English settlements, possession of land was directly related to one's labors for or upon it. Money, capital, allowed for the accumulation of wealth, and its value depended on need. "What would a Man value Ten Thousand or a Hundred Thousand Acres of excellent Land, ready cultivated, and well stocked too with Cattle, in the middle of the in-land Parts of America, where he had no hopes of Commerce with other Parts of the World, to draw Money to him by the Sale of the product?"[4] The system of land ownership in the United States sprang from land's attachment to a marketplace and one's ability to accumulate wealth from the sale of property. The property's care was then assigned to the invisible hand of the owner's self-interest to maintain or enhance his property's value. Society's attempt to incorporate jurisdiction over diffusely occurring resources into the rubric pertaining to surface rights transformed a rational, tidy property system into the morass of the early oil industry.

The United States is often referred to as "a triumph of individualism" and self-determination; Petrolia stands as a demonstration of how these characteristics were allowed to proceed entirely unchecked. The best indicator of the irony of this "triumph" in Petrolia was the abandoned derricks that began littering the region shortly after speculation began in 1860. One observer described these as "decaying monuments of small fortunes ruined when . . . the first oil excitement arose, [and] labor attempted to emancipate itself from capital."[5] In other words, this was the result of the wild speculation of a mass population, with anyone being able to secure leases and sink wells at will.

Searching for oil brings the chance of the gambler to the controlled world of industry. A disgruntled forty-niner would have pulled up his stake and cleared the lease for others, such as Chinese speculators. When the black-jack player folds, he or she leaves behind only cash, now the property of the house. Nothing records the presence of the risks these gamblers take. The oil speculator of the 1860s differed from these players of chance. The oil speculator attempted as many hits as financially possible and then simply bid adieu to the site. Behind him, the oilman left all that he had started—derricks, pump houses, uncapped wells. Anyone could sink a well in the early days of oil. And, indeed, many came and tried.

These boomers found an industry ripe for speculation. The industry of the Oil Creek valley, particularly until 1865, proceeded with only one rule: the rule of capture, which acted as one of the major forces stimulating—not limiting or regulating—massive development. This simple rule placed no jurisdiction of property law over underground crude. The fortune went to the first one to strike and seize the supply, bringing subterranean crude up to Earth's surface where land ownership applied.

The rule of capture spurred a stream of boomers to the valley. Soon every aspect of the industry and life associated with it became based on the immediacy generated by the rule. Whereas the discovery of oil led farmers and others to develop a new industry and technology, the rule of capture catapulted the search for oil to the status of an industrial boom. The rule, in essence, gave the industry the right to have no rules restricting its development.

Crude oil appears inconspicuous enough coming from the ground in a greenish black ooze mixed with water—not the jet of blackness that many of us imagine. Organic wastes, such as plants and microscopic plankton floating in ancient seas, accumulate at the bottom of oceans and lakes. Over millions of years these layers of waste form levels of sediments rich in carbon and hydrogen. Natural degradation converts such residue into hydrocarbons (oil and natural gas) through pressure and underground heat. The droplets of oil migrate through levels of rock before becoming trapped in permeable rocks and sealed in reservoirs by shale rock on top and heavier salt water at the bottom.

Geological magic takes place in these reservoirs. Pressure builds as the matter further decomposes. Added pressure created by the lighter gases in this reservoir over thousands of years allows the oil to flow up the well bore when the drill bit pierces the reservoir. This fugacious quality creates a flowing well: a well of oil out of which a supply of oil flows. While the gusher possesses romantic associations of wealth, it occurs when the pressure is mismanaged. The gusher is wasteful, dangerous, and unnecessary from a practical point of view. This opinion would have been in the minority in the oil boom of the 1860s.

The fugacious quality of oil distinguishes it from other sedentary mineral resources such as coal and wood. The liquidity of the resource forces the application of ownership laws such as those pertaining to landowners' rights to claim water and to run free-roaming animals. Under normal conditions, these legal decisions were fueled less by philosophy than by the possibilities of economic development. Rivers or water wells were the only similar resources litigated during this time period. Interestingly, the U.S. courts consistently supported industrial development in New England over the petitions of farmers whose land had been impacted by the mills' use of the rivers.[6]

With neither legal assistance nor the formal and informal controls of land ownership, what happens to a place quickly dominated by a valuable commodity? Tarbell depicted the Oil Creek valley as overrun by temporary development and exploitation. In relation to other sites of extractive industry, Petrolia uniquely compelled individuals to lose any social or cultural restraint: "Taken as a whole," she stated, "a truer exhibit of what must be expected of men working without other regulation than they voluntarily give themselves is not to be found in our industrial history."[7] Due to the rule of capture, property demarcations could not hold the resource, and therefore the commodity went up for grabs—held in common until the first lucky bit tapped the reservoir. Contemporary writers, such as Garrett Hardin, have extended Tarbell's sensibility to better understand such a locale. Hardin placed such development within an ecological framework by coining the phrase "Tragedy of the Commons."

In his writings, Hardin, a biologist, presents the undeniable effects of human tendencies, specifically greed, on natural resources held in common, such as a fishery, a community pasture, air, or oceans. Human nature, Hardin asserts, as a capitalist society rewards it, will press each occupant to use the resource as quickly as possible so that others do not first reap its benefit. As humans have come to view the environment around them as a shared, global concern, Hardin's argument has been extended to apply to the interconnected commons impacted both locally and distantly by the actions of each human. Within such an argument, land becomes a commons regardless of boundaries such as property lines or political distinctions.

An underground reservoir of crude oil also exists as a common resource. No one owns the oil until it reaches Earth's surface. While surface property rights were adhered to in Petrolia, the lack of restraint created an ethical commons designed to extract the oil from the ground. In the Oil Creek valley, land ownership or even mere lease rights entitled users to administer both the supply and the land in any fashion he or she chose with dramatic impact on the surrounding property and natural environment. Indeed, Hardin seemed to be referring to many of the practices of Petrolia, when he said that "the rational man finds that his share of the cost of the wastes he discharges into the commons is less than the cost of purifying his wastes before releasing them. Since this is true for everyone, we are locked into a system of 'fouling our own nest,' so long as we behave only as independent, rational, free-enterprisers."[8] The early oil industry reified itself by encouraging speculators, wildcatters, and others to behave only as "independent, rational, free-enterprisers." The industry perpetuated itself by creating other oil regions; these, however, would always burn out by design.

For individual speculators, this meant pulling the oil out of the ground as quickly as possible and concerning oneself as little as possible with the condition of the industry's temporary home. The industry reinforced this by celebrating those who best lived up to these ideals and by refusing to place any regulations or limitations on the expansive industry. The ensuing activity indeed left the nest severely fouled.

Not only did the rule of capture spur boomers to the region, it also prodded them to sink wells wildly, with little forethought or planning. Other energy resources necessitated large-scale, multifaceted industrial processes. Coal extraction, for example, followed the organizational pattern of its seam, which was often owned by one corporate extractor. The company extracted the coal methodically: the seam possessed a definable size, and its time for extraction as well as its consumption of labor could be calculated fairly accurately. On the other hand, oil speculators raced to capture any supply before others beat them to it.

British mineral (but not oil) mining law in the mid-nineteenth century established the rule of capture: namely, that the rights of the surface landowner

were practically supreme. The legality of nineteenth-century U.S. oil speculation continued to be tied to the primacy of the surface owner and derived
from the English case, *Acton v. Blundell*, decided in 1843.[9] The original suit
arose when the flow of a percolating water well operating a cotton mill was
depleted in 1837 by a coal pit sunk on neighboring ground. The owner of
the well sued the pit owner and based his claim upon the established English
law of surface streams.

At this time, riparian law held that "each proprietor of the land has a right
to the advantage of the stream flowing in its natural course over his land."[10]
This right, as the ruling stipulated, must not be "inconsistent" with others
living on and using the same stream. In other words, there needed to be an
overseeing authority, most likely a government or court, to administer any
resource such as a stream. This entity would ensure that one party did not
capture more of the resource's energy or ability than another. However, complications grew when one moved beneath Earth's surface.

In *Acton v. Blundell* the British court held that the surface stream doctrine
did not apply to the flow of water as it was drawn from wells. The court's ruling stipulated that the action of the well indeed usurps the liquid from neighboring soil but does not do so "openly in the sight of the neighboring proprietor." Instead, it taps this neighboring flow in "the hidden veins of the
earth," and it would therefore be impossible for any proprietor to know
"what portion of the water is taken from beneath his soil." The court concluded that until this rate of flow could be established, it must assume that a
well usurps no water from neighboring lands; therefore, landowners have
every right to capture as much water as possible by drilling down into their
own property.[11]

The opinion also discussed the serious consequences that were bound to
occur if the law of streams were allowed to take precedence in such a case. If
first sinking a well into a specific stream gave the owner an indefeasible right
to the water in that stream, a neighbor would be unable to make use of the
spring on his own land. In conclusion, the court noted the importance of not
placing limitations on mining when it declared that "a well may be sunk to
supply a cottage, or a drinking-place for cattle; whilst the owner of the adjoining land may be prevented from winning metals and minerals of ines-
timable value . . . there is no limit of space within which the claim of right to
an underground spring can be confined: in the present case the nearest coal-
pit is at the distance of half a mile from the well: it is obvious the law must
equally apply if there is an interval of many miles."[12] The court granted the

"owner of the soil all that lies beneath his surface." This provided the owner with the freedom to use the resources (or hire someone to do so) at his pleasure.

The ruling in *Acton v. Blundell* received its first application to petroleum deposits in the Pennsylvania state courts. These issues, however, were not brought up until well after the Oil Creek valley's boom. The first recorded case to invoke the rule of capture occurred in 1875, and it specifically mentioned that the legislature might wish to create a different law. "No doubt," the deciding judge wrote, "many thousands of dollars have been expended in oil and gas territory that would not have . . . if some rule had existed by which [the resource] could have been drilled."[13] But no rule did exist—except for the rule of capture.[14] In a strikingly apt simile, historian Stanley Clark referred to the rule of capture as "the law of the jungle."[15]

The 1860s boom went on without reconsideration of this law. The laissez-faire approach to development was simply accepted as the standard course. Finally, in 1899, the Pennsylvania Supreme Court substantiated what had been practiced for forty years in the Oil Creek valley, when it ruled that "every landowner or his lessee may locate his wells wherever he pleases, regardless of the interests of others. He may distribute them over the whole farm, or locate them only on one part of it. He may crowd the adjoining farms so as to enable him to draw the oil and gas from them. What, then, can the neighbor do? Nothing; only go and do likewise."[16]

As Tarbell wrote, the earliest speculators in Petrolia did not need to be told to "go and do likewise." When speculators considered the fortune available, a natural human reaction drove them quickly toward sinking a well, and another, and another. Much like playing the lottery, one's chances of a major strike were thought to increase with every well sunk. Hardin's argument comes to the fore as we see how these urges drove oil's early development.[17] But would any of us have acted differently?

With the rule of capture as its main vehicle, the early industry swiftly escalated to boom. The need to seize the resource rapidly heightened the intensity of the boom but did not altogether cause it. Other details of early oil also fed the industry to boom rather than to undergo stable, long-term growth.

The fugacious quality of oil dovetailed with the rule of capture to create a rush of people to the region in 1860. For instance, an October 6, 1859, article observed that "there seems to be no diminution to the supply, and the only difficulty appears to be, to get vessels to contain it until it can be sent to market. Think of 1200 gallons of oil drained from the earth's caverns each twenty-four hours, at an expense of some six dollars, and visions of Pike's Peak will no longer dazzle your vision!"[18] No need to look west to the mountains for opportunity; progress and prosperity are right under your nose. On October 7, the author of the *New York Tribune*'s initial announcement of the oil discovery (apparently Francis Brewer) wrote an account to update all of "those many interested parties" who had written to him since the publication of his September 13 article. Imagine the reaction when potential investors and speculators read these words:

> There seems to be no diminution of the supply however much the speed of the pump is increased. There is land for sale. . . . The nearest railroad point is Union, on the Sunbury and Erie Railroad, twenty-five miles from Erie, Penn., then by stage twenty miles to Titusville. We see many strange faces in our quiet village, and we are happy to see them; at least, I know the hotel keepers are, if I may judge from their pleasant countenances, or the kind attentions to their guests.[19]

Hearing that people had already begun to arrive would have spurred many readers to action, and likely Brewer was entirely cognizant of this fact.

The fantastic wealth possible in oil also attracted many boomers. Some articles presented economic fortune as a simple news event of individuals' increasing worth; others added details that enhanced the feeling of immediacy. Thomas Chase and his wife traveled to Titusville on their honeymoon in 1859 and filed the following report in their local newspaper upon their return.

> As a result consequent upon this discovery, real estate and leases with privilege of boring till oil was found, were each held at great prices. . . . The tract of land on which the large spring has been opened by Mr. Drake, was once purchased by the father of the writer of this article for a cow, and previous to that had been sold at treasurer's sale for taxes. Now we believe $100,000 would not buy one acre of it. Men until now barely able to get a poor living off poor land are made rich beyond their wildest dreaming.[20]

The increase in value of the valley's lands stunned the readers' sensibilities as it did the Chases'. Most observers would assume that they should seize their fortune in Pennsylvania soon or risk prices rising further.

With the rule of capture in place, the primitive technology of the industry and the fashion in which wells occurred also acted as enticements to development. Even when this technology became slightly more advanced in the later 1860s, a type of well known as a gusher overwhelmed the misgivings of any onlooker with visions of being bathed in black gold. With Drake's simple good luck behind it, the industry's first "technology" sought to find some rationale for locating wells. Surface appearance, particularly that of seeping springs, was the most popular method for selecting properties on which to drill. However, after the first few days of speculation, very few such sites remained available.

With physical geography no longer a useful determinant, the industry went metaphysical. The divining rod became the most acceptable method of well location. Using a forked twig from a witch hazel shrub or a peach tree, a diviner would tightly hold the wood as he passed around the property. A swift downward dip of the wood, often imperceptible to the naked eye, would show where one should locate a well.[21] A few successful divining experiences often permanently established a diviner's reputation in the business. Spiritualists and "oil smellers" also made a fine living in the oil fields by locating

Fig. 2.1. Kicking down a well with a spring pole

Fig. 2.2. The iron rod transfer system for dispersing steam power from a prime mover (steam engine) to wells

wells. A spiritualist or smeller often charged a lower rate but was similarly dependent on reputation.

The site secured, speculators hired laborers to carry out the numbing work of drilling the well. The early equipment consisted of universally available items that could be made by any blacksmith. Each well—whether successful or not—carried a $1,000 investment. While a considerable sum, most speculators parceled out start-up fees between a variety of wells and only paid a portion of this investment for any single well. Part of this fee would hire the laborers to actually "kick down" the well.[22] These robust Irish or German men would press a wood beam downward and drive the drill bit into the earth to slowly chip away at the rock, sand, or soil (fig. 2.1). At this point, the typical well came in at around two hundred feet in depth. An experienced team of kickers generated two strokes per minute to sink the bit three to six inches per day.[23]

If this chipping process (inaccurately referred to most often as drilling) revealed a flow of oil, the oilmen attached a pumping apparatus to the wellhead in order to maintain and increase the seep of oil. This derrick mechanism controlled the pump, operating on a vacuum principle, and forced the oil through the tubing and into the wellhead. The pressure created by one centrally located steam engine often simultaneously powered a number of derricks (fig. 2.2). The oil came to the surface mixed with water and therefore needed to be fed into a separation tank from which the water could be drained when separation had been completed naturally.

Striking and operating an oil well required no more than this relatively simple technology (figs. 2.3a,b, and 2.4). Often drillers were blacksmiths, who offered employers the ability to construct or make adjustments in the necessary apparatus in the field. They created or adapted tools to confront each challenge of the early industry. With few known truths or set facts, the oil industry of the early 1860s presented splendid opportunities for men with funds to invest or simply a bit of expertise in industry.

Despite the length of time that it could take to sink a single well, the influx of inexperienced individuals combined with the simple technology and natural occurrence of oil to feed speculators' interest in readily sinking as many wells as possible. Between the locating and sinking of new wells, the influx of new people, and the clatter of the existing industry, it was hard to imagine in late 1860 that the confusion in the valley could worsen; yet the true speculative boom had not even begun.

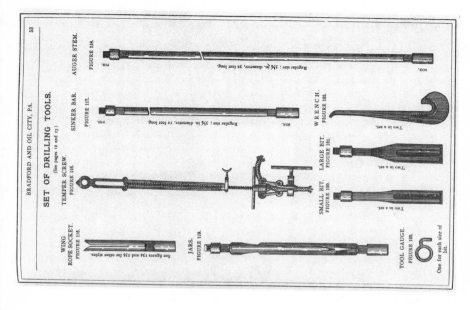

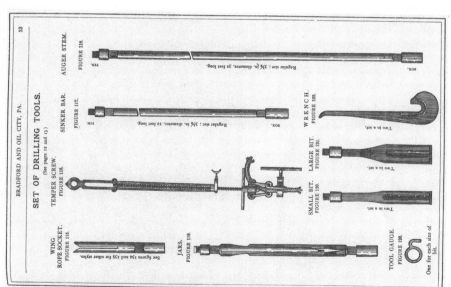

Fig. 2.3a (*left*), fig. 2.3b (*right*). Diagrams of tools, from *Harper's Monthly*

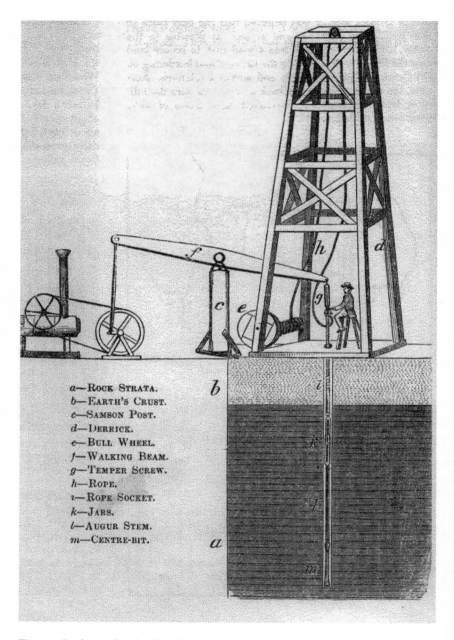

a—Rock Strata.
b—Earth's Crust.
c—Samson Post.
d—Derrick.
e—Bull Wheel.
f—Walking Beam.
g—Temper Screw.
h—Rope.
i—Rope Socket.
k—Jars.
l—Augur Stem.
m—Centre-bit.

Fig. 2.4. Linkage of early oil exploration

While all of this wild speculation intrigued Americans, the labor-intensive process of oil extraction produced only limited amounts of oil, not enough to make it a truly viable source of energy. Certainly, the supply had grown from that seen before 1859, but it remained insufficient to become part of any basic industrial infrastructure. At this point, Americans were attracted to the technology by its oddity and their own curiosity at the economic possibilities if the flow were to remain consistent. Such possibilities were about to be realized in a way that would change everything for the Oil Creek valley.

Through the first two years of development, the strikes were known as pumpers—the derrick was largely responsible for extracting the supply. In April 1861, the unbelievable phenomenon of the flowing well of oil became reality: a well of oil that needn't be pumped; a well in which the oil actually exploded out of the earth. Henry Rouse, frustrated with the minimal production of his well at 150 feet, had set about to drill the well deeper. At a depth of 300 feet the drill struck a pocket of natural gas in addition to a pool of crude oil. The intense pressure of the gas sent the oil spouting 60 feet into the air at a rate of three thousand barrels per day. The boon soon turned into disaster due to the workers' inability to cope with the never-before-seen phenomenon of a flowing well. A fire ignited, killing Rouse and eighteen others while it burned out of control for three days. Although a tragedy had occurred, many oilmen saw instead that the scale of the industry had been completely altered.

In May 1861, the Fountain Well struck at 460 feet. The Empire Well then came in at Funkville and produced initially at a rate of twenty-five hundred barrels per day before continuing after eight months at twelve hundred. In such examples, the names of wells (when not signifying the owner of the lease) reflect their odd occurrence—as if they would remain landmarks in perpetuity. Of these first flowing wells, one observer wrote that the loss of oil had been great: "At first wells were bored with the hope but not with the certainty of oil, and the tank was usually a secondary consideration. When the first [flowing] wells were opened . . . there was little or no tankage ready to receive it, and the oil ran into the creek and flooded the land around the wells until it lay in small ponds. Pits were dug in the ground to receive it, and dams constructed to secure it, yet withal the loss was very great."[24] Speculators felt the expense of constructing tanks at prospective well sites unnecessary until they were certain that the well would produce. However, during the interim many dollars worth of crude would be lost. Many early oilmen based decisions on a cost-benefit analysis founded on the idea that crude was available nowhere else in the world. There existed no competition.

A steady series of deeper, flowing wells put many of the earliest pumper wells out of commission and attracted thousands more speculators to the region. This breakthrough functioned to define the landscape of Petrolia in two ways: the increased impact of outside speculators and the hulking remains of abandoned, less-productive derricks. Even if it produced some oil, a well could be judged not worth the expense if it did not produce at a certain rate. The flowing well, therefore, allowed oil production to cross another plane in the process of commodification: from discovered treasure, the natural resource swiftly became an industrial commodity. The entire landscape would be reconstrued to fit into the process of extracting oil from its geological home. This valley became more a process than a place.

The madly flowing wells furnished new fuel for the popular belief that enough wealth lay beneath the valley's surface to satiate the wildest dreams of many Americans, which then added further to the mythic grandeur of the industry. The *Venango Spectator* discussed the effects of the massive increase in the oil supply from twelve hundred barrels a day in 1860 to over five thousand in 1861, then to astronomical proportions in 1862. "The great depression in the market prices of crude . . . has . . . been the means of introducing the product to all parts of the world and made it as much a necessity as any single article of human want."[25]

Flowing wells combined with overdrilling to convince some Americans that the region's supply would soon be drained—that the supply was finite. This belief spurred many formerly cautious Americans to rush for their opportunity in case the supply were to disappear. Such convictions often were directed by guidebooks for investors, which used faulty geological information to instigate readers to quick action. In this case, the idea of the supply's being overexploited possessed some scientific merit.

One guidebook described the phenomenon of overdrilling in this fashion: "the wells seem to have a more intimate connection, as though the supply of an entire locality was drawn from a reservoir having more or less continuity. Hence, all the flowing wells had their production interfered with, and in most cases stopped, from the sinking of other wells in their immediate vicinity." The flowing wells attracted new speculators to the region, but then also enticed those who were already there to go more furiously about their

work. The same guidebook reported that these "leviathans" puzzled operators of smaller pumping wells. "If it were possible to continue the new mode
of supply, it was argued that the source would soon become exhausted."
Overall, the flow persisted and had become a "permanent boon."[26] As the
Reverend Eaton observed, "Every man on the creek was anxious to have a
flowing well, although the product might [go dry] upon his hands. The dark
green fluid represented wealth; it had made many rich, and large quantities
were desirable in any event."[27]

Similarly, the field technologies themselves very quickly demonstrated
the industry's lack of restraint. While flowing wells had revolutionized
drilling in the valley, the actual sinking of a well remained a time-consuming chore, and one that often did not bear results. Whether a hole went
straight or crooked, it still possessed the same likelihood of coming in a
"duster." Probing blindly with a drill bit only one to two inches in width obviously made striking a well a risky proposition—particularly when a spectator was interested in reaching every single pool occurring beneath him.

Most oilmen considered the technology that could effectively bring crude
out of the ground and to market the greatest problem facing Petrolia's development; therefore the tasks to which technological innovation should be
applied were those most closely related to industrial production, not to efficiency. The best example of this thinking arrived in January 1865 in the form
of another colonel—this time a real one. Colonel E. A. L. Roberts had fought
in the Civil War, and he knew explosives. With him he brought to the valley
half a dozen torpedoes. These first explosives were cast-iron flasks, filled with
gunpowder and ignited by a weight that dropped along a suspension wire
onto percussion caps in the flask, all of which could be located deep below
the surface.[28] Soon, the technology would evolve to utilize nitroglycerin, a
more volatile but manageable explosive. There appears to have been no out-
cry about the application of such explosives in a region where smoking had
been outlawed and lamps were only used indoors. The industry's new standard of development became detonating a torpedo of nitroglycerin underground in proximity to another flammable mass in hopes of widening the
area from which the well would draw. In addition to the danger at the point
of impact, the transport and rigging of such explosives presented serious danger; yet, these considerations had no standing in Petrolia.

On January 28, 1865, Roberts successfully discharged two of his eight-
pound torpedoes into a well on Watson Flats, near Titusville. There was no

jet of oil produced; but once the debris was cleared, a well that had been slowly petering out now emitted a steady flow.[29] Roberts set up his company in February and drew up his fee rate of $100 to $200 per charge and a royalty of one-fifteenth of the increased flow of oil. Soon, many wells stood as examples of torpedoing's magical abilities to transform a dry hole into a producer. Early in 1866, for instance, wells on the Tarr Farm that were pumping only 3 barrels a day were torpedoed and became flowing wells producing 80–180 barrels each day.[30]

The results could not have been more invigorating to onlookers. The explosive's detonation would often cause a great jet of oil and debris to rise straight out of the hole and high into the sky. Historians Harold Williamson and Arnold Daum, however, observed that "the introductory years of torpedoing exacted a heavy toll in lives, both in factories and in the field."[31] By 1870, torpedo technology had become standard practice, even though the necessary use of nitroglycerin led to accidents. A technological process such as torpedoing, which allowed the industry to create the surplus to support its boom, was worth a certain amount of ancillary costs.

The air of urgency that the rule of capture blew through this valley did not only alter the technological processes of the young industry. The pattern of ownership and risk also greatly shifted from that of the earliest days when the boom revolved around individuals developing their own leases. By standardizing long-distance investment and lessening financial risk, the oil boom swiftly defined a new age in land ownership and use rights. Financial investment and land use employed the lease as its major organizational tool.

The lease, of course, functions as an agreement between the landowner and another party who wishes to use the property in some manner. The agreement secures remuneration for the landowner as well as some control or restrictions over the way the site is administered. Most landowners received a royalty payment varying from one-eighth to one-half of the oil production. If the prospects seemed particularly favorable, the lessee also paid an immediate monetary bonus. With such financial possibilities, only a foolish landowner sold his or her land in the Oil Creek valley during the early 1860s. Land values repeatedly rose beyond anyone's imagination. It was impossible to foresee the next development in the economic boom. Instead, most owners leased their land to oil speculators and refused to sell until later in the 1860s.

While they do not explain the rationale of residents, maps support the continuity in ownership along Oil Creek between the Allegheny River and Titusville from 1857 to 1865, as we shall see in Chapter 5. Interestingly, the landscape of Petrolia suggests that stewardship tendencies were enhanced little by this continuity. Therefore, instead of rapid changes in land ownership leading to exploitation, the significant transactions of this early era revolved around the trading, selling, and swindling of leases. Joint stock companies drove this economic period by purchasing leases from landowners. These leases would then become the main gambits in the process of "oil speculation." Companies bought and sold shares in leases at opportune times. Most important, the joint stock companies parceled the leases into smaller shares, thereby dispersing the risk among many lessees and investors.

Beginning in 1860, landowners and stock companies purchasing leases peddled them for an acre or less of ground. For such a lot, they would accept a money payment for the lease rights and would also establish a rate of royalty interest. During the 1860s, the Oil Creek valley was cut up much like a sheet cake, but then swapped and sold many times before anyone took a bite. The frenzy of leasing surface rights and the practice of such land speculation reflected the commodification and ownership of land gone almost crazy.

With the financial risk dispersed so widely, so too went any commitment to permanent life in this region. The confused morass of lease trading and abandoned wells left landowners with almost no ability to monitor or control the actual activities carried out on their land. Oilmen were free to develop the valley as they wished, with little knowledge of what had gone on in this place previous to the oil industry. In addition, the specific decisions concerning land use were left to businessmen, industrialists, and laborers who would move on with the industry when Petrolia's supply had run dry.

By the time land users took up a plot, the land had become distant from its cultural meaning and ecological significance. It had been transformed into a commodity, much like a box to be emptied. By 1870, the lease even conveyed these principles by stipulating the number of wells to be drilled within a definite time limit and containing clauses of forfeiture for failure to carry out the agreement. These conditions impelled a producer to drill wells when the market price did not warrant the expense. Lessees often overdrilled to keep from forfeiting a lease in which they may have already made a valuable investment or in the belief that oil's value would soon rise.

Leases also spelled out other ways in which holding companies conducted land use in a shortsighted manner. For instance, the length of leases varied from twenty to ninety-nine years, and some extended in perpetuity. A minimum drilling depth would be stipulated so that the lessee neither gave up too easily nor stopped because of too light a flow. Frequently, two hundred feet stood as the minimum drilling depth, although in 1860 a full five hundred feet occurred regularly.

During 1860 and 1861, the royalty rate ranged between one-quarter and one-third of the oil produced. Usually, drillers delivered barrels of a site's crude directly to the landowner. Cash outlays were often stipulated and paid before drilling or when production reached a certain level. Such arrangements, of course, functioned to entice landowners not only to lease their land to others but to create as many parcels as possible. Suddenly, economic motives made it highly desirable to have representatives from all over the nation acting as the stewards of one's property.

The trading also did not stop at the trading company's level. Subleasing accentuated almost all of the problems of the existing trade network. Subleasing involves the selling of fractional shares by lessees, thereby further distancing the user of land from its owner. The holder of a lease for ten acres could sublease all ten acres if he chose and never sink a well himself. Or he could develop the entire ten acres but accept investment support that he then rewarded with partial shares in his lease. Subleases made the oil industry the domain of speculative specialists, who acquired leases and subdivided them at huge profits to producers. To smaller speculators, subleasing afforded the means by which they could bolster resources and expand their drilling. Subleasing also helped to spread the risk if one had few resources. By 1860 a speculative infrastructure was in place that distinguished the oil industry from any other. One could be in the oil business and make a fortune at it without ever getting grease on one's hands.

Such trading of any sort occurred through personal contact at the well, boardinghouses, and hotels, or by chance meetings on the streets of Titusville, Oil City, or other eastern cities. One such locale, along the elevated wooden sidewalk on Centre Street in Oil City, was called the "curbstone exchange." However, after the railroad had connected Titusville with Oil City in 1866, many of the deals were struck in a special coach specifically designed to serve as a meeting place for producers, speculators, and dealers in the industry. The informality of the trading defined the early industry; however, as oil became big money, many traders pressed for more organized exchanges. The first Oil Exchange was formed in Titusville in 1871, and others followed in Oil City and Franklin.

The trading of the oil industry did not remain a secret for long. The traders soon were backed by capital from all over the world. The long-distance investment of blind capital took the valley one additional step away from landowners controlling their land. The oil industry was the first extractive industry—in many ways the first land-use practice—to draw from long-distance financial speculation.[32] Advertisements for investors began to appear in the *New York Times* in the early 1860s. While the Civil War tempered the possibility for investment, the postwar years presented a flurry of economic activity in the valley. This type of investment provided speculators and companies with nearly unlimited resources with which to exploit oil supplies. The only true limit on earnings was the amount of land and oil available in the Oil Creek valley. Much as in a foreign-backed war, this region became site to an industrial undertaking that had neither understanding nor appreciation of the local culture and ecology. In fact, most investors did not even know where this place really was.

These characteristics of the oil industry, which basically stem from the rule of capture, distinguished it from any previous industrial undertaking. A modern reader is forced to shake his or her head and ask a simple question: "Who actually *owned* the land?" The answer is simple: by 1864–65, most often larger holding companies or absentee owners who had left the area. However, contemporary readers mean more than who held the property rights; they are really asking, "Who cared for or about the land?"

The answer leads to the ugly truth of the rule of capture's effect on oil speculation. As the land became more and more removed from its original owner and steward, any ethical consideration of its treatment grew less and less primary. With each traded leasehold right, it appears that the land incrementally lessened in its standing for any reason other than producing oil; concurrently, the tract of land's value as property would rise with its association with the product being taken from it. In other words, the landscape itself would sink deeper and deeper into the process of commodification that was infecting the entire Oil Creek valley.

Drake himself was oblivious to the attention being paid to oil's discovery. He made no effort to secure landholdings or leases, nor to patent any of the processes he had employed. His employer, the Seneca Oil Company, dissolved in 1860, and its leases were taken up by the Pennsylvania Rock Oil

Company. While the industry that he had started flourished throughout the 1860s, Drake was left with no personal landholdings, nor any revenue from wells.[33]

The pen-toting rider, Jonathan Watson, on the other hand, quickly realized what Drake's hard work could produce. Due to his foresight, Watson secured options on many properties, including the Hamilton McClintock Farm, which housed the most famous oil spring along the river. He immediately placed a well on it, the Barnsdall Well, which lay just north of Drake's site. By February, the Barnsdall had reached 112 feet in depth and pumped fifty barrels a day. The Crossley, the third major well undertaken, produced seventy-five barrels a day by March. An account from the *Jamestown* (New York) *Journal* indicates the emotions generated by each of these wells:

> The greatest excitement exists in [this] region, and fortunes are made in a few minutes by sale or lease of lands. [One speculator] bought 300 acres for $30 per acre, and then leased it for $300 per acre, and ¼ of the oil found. Wells are sinking in every direction, and strangers are flocking in from all parts of the country.[34]

George Bissell described the scene of early speculation in his November letters to his wife:

> We find here an excitement unparalleled. The whole population are crazy almost. Farms that could have been bought for a trifle 4 months ago, now readily command $200 and $300 an acre, and that too when not a drop of oil has ever been discovered on them. So much for the bare hope of their being by any possibility a sub-stratum of oil. . . . No California Placer was ever one tenth part so valuable. When the other springs are opened the profit will be millions. I never saw such excitement. The whole western country are thronging here and fabulous prices are offered for lands in the vicinity where there is a prospect of getting oil.[35]

The idea of skyrocketing land prices drove those only vaguely interested in investing to go to Petrolia sooner rather than later. Bissell himself was not far behind Watson's ride. Four days after the ride, Bissell began leasing or buying farms until he had expended $200,000 by the end of autumn.[36] Other prospectors arrived from Pennsylvania, New York, Ohio, and New England.

These were only the precursors of a mad and maddening rush that lasted a dozen years. Investors sunk incredible sums into the real estate of this eleven-mile valley, and the amounts would rapidly be dwarfed again and again. The liquid nature of petroleum allowed it to fall between the cracks of American ownership law. The scene, then, became a dramatic illustration of the formal and informal controls that ownership can make possible, with the tragedy most affecting the region's common ecology and culture.

Whether speculators constructed a spread of derricks to fulfill their lease agreement or to drain a pool of oil before anyone else, their motives were the same. As Jonathan Watson's horse galloped along Oil Creek in 1859, the fate of this valley fell into distant hands. Outsiders would have more to say about the use of this place over the next century than any locals. Boomtowns would dominate the region, and the cities that existed previously, such as Titusville and Franklin, would be permanently altered. Oil flooded out the meaning of local culture as it also usurped the values with which the valley's resources were judged.

These changes in inhabitants' use of the natural environment make the Oil Creek valley of the 1860s a unique stage on which to assess and understand the individual drive for success and wealth. The residents' values and motives reveal themselves in the technological developments carried out to solve problems and fulfill priorities within the evolving industry. In addition, the priorities reveal themselves through the lack of any effort to limit or regulate speculation.

All of these land-use tendencies enable us to understand and reconstruct the ethics and values regarding this valley's use of its natural resources during the 1860s. As the ruling justice observed in the Pennsylvania Supreme Court's 1875 decision concerning the rule of capture: "The discovery of petroleum led to new forms of leasing lands. Its fugitive and wandering existence within the limits of a particular tract was uncertain, and assumed certainty only by actual development founded upon experiment. The surface required was often small compared with the results when attended with success, while these results led to great speculation, by means of leases covering the lands of a neighborhood like a swarm of locusts."[37]

The swarm of locusts followed Watson's ride and left the neighborhood soiled as no place else. Yet the system functioned just as it was supposed to. The landscape of Petrolia demonstrated the tendencies of the American system of land ownership when a region is allowed to boom with no regulation—with no laws but the rule of capture.

*In scale and desolation — and, I am
afraid, in duration, [such] industrial
vandalism . . . has no human scale. It is
a geologic upheaval. . . . And the ruin
of human life and possibility is commensurate with the ruin of the land. It is a
scene from the Book of Revelation. It is
a domestic Vietnam.*
—Wendell Berry, "Mayhem in Industrial
Paradise"

Chapter Three

The Sacrificial Landscape of Petrolia

The creek, sinuous as a snake, can be seen for three or four miles, its flat banks
covered with derricks, vats and engine houses that jostle each other in confusion, while glimpses of the bends beyond can be obtained, equally full of life
and business. The creak and wheeze of the engine and the pump are mingled
with the clank of the blacksmith's anvil, and the shouts of the flatboatsman
urging his team of four horses abreast up the middle of the stream with a load
of barrels. Looking down on such a panorama of busy life . . . it is impossible
not to be struck with the immensity of the business which has but recently
sprung into existence.[1]

Perched atop one of the hills surrounding Oil Creek, the journalist writing
this quote did not need to mention the commodity of which he spoke. The
reader knew the topic as oil and the locale as western Pennsylvania. The one-
ness of product and place in this valley made such inference possible and often inevitable. The scene was typical of business, enterprise, and economic

progress; however, this passage paints a physical landscape—not an industrial process. Before the pump houses, derricks, and boardinghouses, a place of importance had existed here. Even after the oil boom, this place peeked out from the development overwhelming it.

A landscape is constructed of geology, hydrology, and biology; yet it also includes the creations of the humans or other beings that inhabit and change the environment. Where nature and culture meet, they construct a landscape.[2] This construction is most obvious in its physical manifestations, yet humans also determine its spiritual, social, and cultural meanings. Therefore, such a meeting between nature and culture may not always result in a physical creation. A vision of a place can also form within the mind, as humans reshape attitudes and values—thereby adding a mythic component to the meaning of an envisioned locale.[3] In this fashion, a definition of place can be constructed externally by a larger culture. Occupants may still form their own ideas of a place, but an external construct based in ideals of the larger culture also encroaches on a place's meaning.[4]

In the case of Petrolia, the first details to come to the public mind were complex combinations of actual events or practices mixed with mores or views of the national popular culture. While the cultural significance of the Oil Creek valley derived primarily from the activities of the 1860s oil industry, the valley of the 1860s also became at once a wonderland of financial opportunity and a harrowing invitation to disaster. The American public invested the Oil Creek valley of the 1860s with a cultural significance that extended well beyond the Allegheny Mountains. The fortune and danger, neither of which had local roots, defined the region's meaning and future.

If industry had proceeded in this locale as it had in others, the valley would have neither boomed nor proven a watershed in the history of American industry. The technological development would have simply functioned to devise the most efficient and productive method possible for extracting, transporting, and processing crude oil. But in the case of the Oil Creek valley, external culture and outside investors overwhelmed the sleepy region in an effort to fulfill all the dreams that were quickly being tied to black gold.

The setting of these dreams and stories became critical. No matter its specific topic, popular coverage of the industry first set out to relay the stunning details of Petrolia. The more spectacular and base the scene, the more intriguing to readers. No detail appealed like the landscape of oil. Such coverage functioned to worsen and even broaden the impact of laissez-faire practices by rationalizing them and making them appear not only acceptable but also progressive. As the property system functioned to escalate this industry and place to boom, the cultural obsession with technological growth threw fuel on the fire and made a myth.

In a related by-product of this depiction, journalists' portrayals of the scalded and smothered natural environment of the Oil Creek valley established a new distance between humans and the natural environment. Writers portrayed the Oil Creek valley as a place where environmental ethics of any sort were secondary to the generation of wealth. Descriptions carefully stipulated that the pollution and waste considered base elsewhere were signs of progress here. Through such portrayals, the landscape of the Oil Creek valley became one of the nation's earliest sacrificial regions—the vanguard of sprawling refinery-scapes, toxic waste dumps, and the coal strip mines so prevalent a century later.

Petroleum's flammability captured the reader's attention more than any other detail. The attempt to bring oil safely from Earth into a barrel without damaging the communities so near the derricks proved the challenge that most titillated readers. This degree of danger came as near as the industry could to six men in a dinghy chasing the largest mammal on Earth. Similar to the whaling tales, the successful coexistence of men and flammable oil intrigued readers worldwide; however, unlike tales of the sea, failed attempts at this coexistence also held great appeal—maybe greater. The experience of H. R. Rouse, mentioned earlier, demonstrates the point better than any other.

In 1861 Rouse had the good fortune to bring in one of the industry's first gushers; however, neither he nor the industry had yet devised any method for handling massive quantities of crude. In the past, there had been ample opportunity to collect barrels of percolating crude after the strike had become a certainty. With a well flowing out three thousand barrels a day, no time could be taken to contain the oil or to take precautions in the surrounding vicinity.

Within moments, a terrific explosion took place, setting on fire all of the surrounding land and buildings, as well as the people who had gathered to view the gusher. Rouse's clerk, George H. Dimick, offered his recollections of the incident nearly a decade afterward. "On the instant," he recalls,

> an acre of ground with two wells and their tankage, a barn and a large number of barrels of oil were in flames, and from the circumference of this circle

of fire could be seen the unfortunate lookers-on of a moment before, rushing out, enveloped in a sheet of flame which extended far above their heads, and which was fed by the oil thrown upon their clothing by the explosion. . . . So numerous were the victims of this fire and so conspicuous, as they rushed out, enveloped in flame, that it would not be exaggeration to compare them to a rapid succession of shots from an immense Roman candle.[5]

According to the *Eclectic Magazine*, "a *jet d'eau* (so to speak) of oil caught fire!" Not only did the flowing oil well catch fire, but also the gas emanating from the well:

In a moment the whole air was in roaring flames. As soon as the gas took fire, the head of the jet of oil was in a furious blaze, and falling like water from a fountain over a space one hundred feet in diameter, each drop of oil came down a blazing globe of boiling oil. Instantly the ground was in flame, constantly increased and augmented by the falling oil.

Such a description would seem to be simply a statement of the necessary horrors of this new product; yet the coverage of fire then reveals something about American readers in general during the 1860s. The writer vividly portrayed the horror for every reader by coming to rest on the experience of Rouse, who was burned in the fire:

When he arrived, not a vestige of clothing was left upon him but his stockings and boots. His hair was burned off, as well as his fingernails, his ears, and his eyelids, while the balls of his eyes were crisped up to nothing.[6]

In the awful days of early oil, Rouse remained one of the good guys. He merited respect in life and death. However, the popular press knew nothing of him except the details of his widely observed demise. Such details stoked the fires of the myth of Petrolia, which separated the oil industry from other extractive undertakings.

Today, Rouseville, named after the deceased, endures as the site of the remaining refinery in the region. It reminds most residents of the region's past and the trickle that defines its present and future. However, placed in the context of the region's mythic past, the site becomes a symbol of the neglect within the industry that cost men like Rouse their lives. It also reveals the odd American attraction to reading about it, which only grew more insatiable as writers tried to top the oddity of their last report from Petrolia. Within this transmission of events a place lost its soul, sacrificing meaning for a brief gasp of economic fortune.

During the 1860s, the popular media made sure that the Oil Creek valley meant oil and that oil meant this valley: they were one and the same. Published articles stand as cultural artifacts of the information reaching the American public of the 1860s concerning petroleum and the regions producing it.[7] The wide variety of magazines covering Petrolia suggests the breadth of the spectacle's appeal, but their presentation also illustrates the wide variety of readers who were interested in the early oil industry.

The forms of this coverage often followed typologies that can be categorized into pseudoliterary genres, including travelogues; narratives of divine progression, nationalism, fortune, and business; and gothic or horror tales. Yet the appearance of such a place in any of these cultural realms could not occur without Petrolia's first being recast as a mythic locale. From this starting point, each of these genre articles contributes to extending Petrolia's mythic status. Made up of these various components, Petrolia emerges from being merely an industrial site along Oil Creek to one that stirred the interest of Americans.

The industry's success depended on myth. Nearly every boomer coming to the Oil Creek valley did so with a great deal of excitement and exuberance; if they hadn't, they most likely would not have persevered. For this reason, nearly every written description of this place also involves such emotion. The extremely positive or negative emotions associated with the industry even influenced the facts of the early years. The dominance of these emotions in written accounts allows the facts to construct not so much a history as a cultural myth of the place.

Cultures often utilize myths to explain the past or the origin of something. In this instance, the need to explain the birth of the oil industry partly drove mythmaking. However, mythmaking also looks in the other direction to explain the future. Cultural historian Warren Susman has written that myths unify the whole, answer the largely emotional needs of members of the community, and provide "the collective dreams of the society about the past, the present, and the future in the same instant."[8] During the 1860s, the popular media placed the oil industry directly in the past, present, and future of the nation, thereby fostering its presence in many Americans' aspirations and dreams. In the process, the myth surrounding the oil industry redefined American ideas about the use of natural resources.

The importance of these popular accounts lies in their effect on the readers of the 1860s. For instance, the *New York Times* journalist quoted at the beginning of this chapter went on to convey his own feelings toward the sites in the valley before him: "My pulse beats as temperately as that of most men; I had neither territory to sell or to buy, nor did I expect to invest much or little . . . in the seductive fluid, yet I must confess to a little tingling in the blood as I contemplated the immense business beneath me, sprung out of nothing, and thought of the many poor men . . . who had grown suddenly rich in that muddy and greasy valley."[9] The seductive effect of accounts such as his resembled his own reaction to the enterprise. Through the popular media, facts about the region combined with hopes of individual wealth and the potential for petroleum to dramatically improve American society. Together, these dynamics formed myth. As the industry evolved, this myth became less prominent; however, the intense interest stirred by the myth was crucial to the development of the oil industry.

At the heart of the myth of early oil and the correspondents' descriptions lay a landscape in cultural transition from agriculture and small-scale industry to massive industrialization. Journalists created headings such as "Oildom," "Oildorado," and "Petrolia" to include each aspect of life in the Oil Creek valley that went into making up the myth of the oil industry. Those living here did not plant and nurture a crop; they violently interacted with the natural environment in order to create two equally undeniable end products: potential material wealth and a drastic marring of the natural landscape. Each became integral portions of the region's mythic status.

Travelogue-style articles first presented Petrolia as an exotic land of buried treasure. Constructed around one writer's trip to and through the region, these articles surely influenced interest and thereby investment in the region's burgeoning oil companies; however, they rarely contained any specific mention of ways to invest in the business. In their accounts, visiting writers first attempted to find words to offer readers a description of this odd place. While such lurid details would certainly pique reader interest (and sell papers), these regional descriptions also allowed writers to express their own astonishment. For instance, this description of Oil City from an 1865 article in *Harper's New Monthly Magazine* titled "After Petroleum":

> I landed. Oh! is this Oil City? Whew, what smells so? Nothing but the gaseous wealth of the oily region. But pigs, mud, no sidewalks! Ah, but you are on the bank yet. Business can not afford to wash the ways down which oil barrels run, nor to scrub their leaky sides. Wait until you reach the main thoroughfare, the grand promenade, the fashionable street of the place. I waited. That is, I walked between wells and oil yards, barns and pens, along the slippery way,

keeping my bearings as I could. . . . Up one side rose a ledge of shale rocks, crowned on top with the "primeval forest." At its immediate foot ran the street. No it didn't run. It couldn't run. Neither could it stand still. It was just too thick for water and wholly too thin for land.[10]

The landscape of this description is the abused, moving landscape making up most of Petrolia during its boom. Only well above the industrial landscape can wild nature, "primeval forest," survive as a witness to all that goes on below.

The *Harper's* correspondent seeks to create the familiar perspective of a speculator traveling to the region for the first time. His trip becomes odyssey as he feels his way through this remote, surprisingly uncivilized region, which he makes appear even more remote and exotic by employing modes of transportation ranging from train to horse, river skiff, and foot. As the writer moves along the river in the skiff, abandoned derricks begin to litter the landscape. These he calls decaying "monuments" of many small fortunes ruined.[11]

Men with few dollars, the writer reports, came to the region to sink wells and attempt to make "splendid gains that generally fall to the rich man's share." The wildcatters leased farmland with little if any money up front. Most agreements hinged on payment to landowners of a portion of the oil money, therefore linking landowners and wildcatters in the promise of the young industry. Speculators then proceeded to sink their wells with no "order, system, or forethought." "Now, through the whole oil region abandoned derricks stand rotting slowly down, warning many and attracting more." Whether warning or attracting, the "monuments" intrigued all, especially far-off readers.

The *Harper's* article continues its travelogue motif by giving very specific descriptions of towns and wells. However, in doing so it merges with other genres in its revealing description of the Oil City area.

The soil is black, being saturated with waste petroleum. The engine-houses, pumps, and tanks are black, with the smoke and soot of the coal-fires which raise the steam to drive the wells. The shanties—for there is scarcely a house in the whole seven miles of oil territory along the creek—are black. The men that work among the barrels, machinery, tanks, and teams are white men blackened. . . . Even the trees, which timidly clung to the sides of the bluffs, wore the universal sooty covering. Their very leaves were black. Only up toward the sky under the clouds, away along the tops of rocks, could the verdure of nature be seen. All below was in sombre clothing, except the sparkling creek, which rattled along its shallow bed. This in the sunlight glittered not like a silver thread in a setting of jet . . . but in fact like one fast moving string of opal. The wasted oil from numberless wells floated over its whole surface, and spreading thinly out from bank to bank, reflected in changing beauty, from every ripple and wave.[12]

The overwhelmed writer presents a vision of the pastoral landscape utterly transformed by the industrial development taking place. In an extension of the sublime literary genre, natural beauty remains above. Similar to the technological sublime, this "pristine" nature seems to appear even more beautiful because of its contrast with the tainted Petrolia. The author heightens the impact of his description by carefully cataloging each detail of the scene before jarring the reader by stamping each detail "black."

The nineteenth-century reader would have been as jolted as are modern Americans at the sight of an oil-soaked seal or water bird; yet the account contains no remorse. Similarly, the artwork accompanying the article is not overtly repellent: lithographs show derricks that appear more like steeples in a small American town straight out of *Picturesque America*. No one questions whether the landscape should look any other way. This was the image left with the reader—the soot as integral and exciting a portion of life in Petrolia as the oil itself.

Just as the literature presented the landscape of early oil as inevitable, the entire industry also became the irrefragable next step in humans' economic progress. Often, the writers' impulse perpetuated great feelings of nationalism. The oil industry became part of a narrative of divine progression—a story of American "chosenness." The divine progression motif became part of a "new religious style," which sought to use popular literature to bring religion to a wider audience by applying it to contemporary issues.[13]

Harper's employed such a mode when it returned to the popular topic later in 1865 with an article opening with a discussion of the gold rush, moving to the discovery of precious coal. The image of coal was extended into an economic fairy tale of developing energy resources and fulfilled needs. "Seemingly not satisfied with the present developments of mineral wealth bestowed on us," the author wrote, "nature, keeping pace with the necessities of man, suddenly unfolds another wonder—*oil, Petroleum*—which now comes spouting from the bowels of the earth, from inexhaustible basins hidden deep down amidst the sandstone rocks below."[14]

The description continued with nature controlling human destiny; nature selected the nineteenth century in which to develop petroleum, which existed beneath Earth's surface for centuries. Just as whale oil supplies were expiring—about to become "among the things that were"—a never-failing supply of burning oil emerged in the mountains of Pennsylvania. The writer blamed nature for allowing the supply of whales to diminish, not greedy humans for overexploiting Earth's mammal resource. Consequently, the emergence of rock oil derived from a natural plan as well, and not the carefully plotted, financed, and carried out experimentation and exploration of capitalist investment.

Once again, the photos and drawings accompanying the text make the towns of Petrolia appear quaint. The scenes in these lithographs demonstrate an interest in fitting this place into a certain desirable type of place. These images contrast dramatically with the photos of the same period taken by John Mather, which reveal the waste and grime of a true industrial site. It appears that magazines such as *Harper's* were not under any constraints in the writing they could publish, but still had to present visual images that were appropriate for "coffee-table" periodicals.

According to the writer, this period culminated in the striking of the Drake Well on August 29, 1859. "Almost in a night a wilderness of derricks sprang up and covered the entire bottom lands of Oil Creek." The entire population of the region had to consider the awesome wealth available. "Merchants abandoned their storehouses, farmers dropped their plows, lawyers deserted their offices, and preachers their pulpits. The entire western part of the State went wild with excitement."[15] Boom had come to the valley, and the nation, state, and region were no doubt better for it.

If one can suggest that the popular consciousness also possesses a subconscious mind, the divine progression narrative most likely grew out of a need to qualify the massive change of this era. Whether for this or other reasons, the effort to assuage any national guilt for this economic boom is very obvious in the popular oil literature. The first of these articles, which appeared in 1862, sought to demonstrate petroleum's long history prior to Americans' making it a valued commodity. In dramatic fashion, these facts built up readers' pride in the American industrial spirit by making this spirit the logical impetus that distinguished oil in the United States from its prior existence in other countries. These other nations and cultures had not overlooked this resource; Americans simply possessed the unique ability to create valuable commodities from unappreciated resources.

The contemporary role of American industrialists in this divine progression of the human species inevitably led to great feelings of nationalism. In "Petroleum, Old and New," *Merchants' Magazine* explained the past documentation of oil pools throughout world history, particularly during the greatest empires of human civilization.[16] This connection between nationalism and divine progress was also demonstrated by the articles cited earlier, which viewed the oil discovery as a reward for the nation's great suffering during the Civil War.

In this genre, a lack of scientific understanding of oil wells often enhanced the mysteriousness of the endeavor. A writer for *Merchants'* explains that one critic of the new industry had forecast that the "world will soon stop revolving for want of something to lubricate its axis." The writer suggested that such pessimism (but not the notion of Earth's failing to spin owing to a lack of proper lubrication) was ridiculous; instead, he believed that "we have ascended another round in the ladder of Progress." The article mentioned some "pious souls" who see this development as the beginning of the world's biblical end—in fire. "Even if their theory were correct," the author said, "there would be no terror in it, for when the appointed time comes, the world might as well burn quickly as to be long about it."[17]

Articles from *Continental Monthly* and *Every Saturday* also smack of nationalist fervor by discussing technological breakthroughs in the past and America's relative position of greatness following its perfection of the oil industry.[18] Such a claim, however, functioned most effectively with a foil—a nation that had not realized petroleum's potential. One 1866 article gave this distinction to Canada, whose oil wells had developed at a much slower pace. The article mentioned that the wells were found soon after those in Petrolia; however, "the lack of energy, enterprise and skill, which is characteristic of the Canadians, marked the history of their oil development."[19]

Just as the development of Petrolia meant economic progress, so too did it mean ways of life being lost or at least altered. Writers often delicately tried to present the wonderful changes for the better while also saluting or at least mentioning what had disappeared. The popular coverage contains a few articles or portions of articles that can clearly be placed in a genre best called nostalgia. Most often, the nostalgia was a longing for simple agricultural ways of life that had become no longer necessary.

The 1865 article in *Harper's* clearly presents evidence of the genres already discussed; however, it also contains a passage teeming with nostalgic reverence for the region's past. "Rich farms are laid waste. The plow turns no more furrows, the scythe cuts no more bending grain. The farmer's barns are no more loaded down with fruitful harvest. The farmer himself, with his homespun clothes, is seen no more in the fields. All is changed! The farm is sold! The old man and his grown-up sons are worth millions, and the old homestead is deserted forever."[20] The author could not conceal a note of respectful longing as he considered the fading pastoral symbols. However, he was not judgmental. Here, humans' use of the landscape, including the abuses made necessary by industrialization, were grouped with the original discovery of oil as a part of God's and nature's plan for inevitable human progress.

Journalists' nostalgia often mirrored that of local residents in that any remorse for lost ways proved fleeting. In the *Harper's* account the author slid from that which had been lost to the commerce and industry gained through the development of a boomtown. In 1857, the correspondent wrote, Titusville had one hundred fifty inhabitants, about thirty-five buildings, and lumber worth $5–$10 per thousand feet. There were no imports, only exports. By 1865, the population had grown to six thousand, with over one thousand buildings, carrying the price of lumber over $50 per thousand feet.[21] How could any American read the description of this change and not lust for the same for his or her community?

These dreams of personal fortune also made up their own genre of Petrolia's press coverage. Particularly due to the broadening economic involvement of long-distance speculation, the actual process of making money in oil became legendary—especially as a lucky few came away with unbelievable fortunes. The national interest was piqued by popular songs, such as "Famous Oil Firms," and even poems such as Samuel C. T. Dodd's parody of Lord Byron's "Isles of Greece," which begins:

The land of Grease; the land of Grease!
Where burning Oil is loved and sung;
Where flourish the arts of sale and lease
Where Rouseville rose and Tarville sprung;
Eternal summer gilds them not
But oil wells render dear such spot.

The ceaseless tap, tap of the tools,
The engine's puff; the pump's dull creak,

The horsemen splashing through the pools
Of greasy mud along the Creek,
Are sounds which cannot be suppressed
In these dear Ile-lands of the Blessed.[22]

Such poems and songs built up the myth of Petrolia, which thereby intensi-fied interest and intrigue, resulting in more capital to further the industry's development. Clearly the waste and marred landscape became a requisite of the process of gaining oil, as did head-spinning investment, clattering iron tools, and puffing steam engines.

So pervasive became the Petrolia myth that often the nuts-and-bolts cov-erage of the business would connect with popular productions like oil-boom tunes. In 1864 the *New York Times* proclaimed that a virulent disease known as "oil on the brain" had become epidemic, "affecting all ranks and condi-tion of men, and spreading rapidly from city to city, until now the whole country burns with oil fever."[23] The use of this phrase demonstrates the wide impact of the Petrolia myth—"oil on the brain" was already a recognized phrase prior to the release of the well-known song. These articles seemed to assume that popular interest would be transferred into investment.[24]

Surprisingly, even the awful details of Petrolia's landscape spurred eco-nomic interest. In the same *New York Times* article, the writer's first lines, placed like a trophy atop an imagined mantle, announced that he had sur-vived two weeks of roughing it in "the dirtiest, greasiest, muddiest, and busiest patch of Uncle Sam's domain."[25] He immediately proclaimed that it was not until his trip to Petrolia that he was finally able to understand the saying "as rich as mud." To this writer, costs to the landscape and financial gain necessarily coexisted. Undoubtedly, the most basic fact concerning Petrolia was that there was money to be made. But this was the working per-son's wealth—one where you got down and dirty and risked everything just for *a chance* at a gusher.

The mythic components of Petrolia did not persist at the same rate as the supply of crude. Great fortunes made in oil continue to hold attention today, but more serious-minded investors in the early 1860s sought out popular writing that could cut through some of the myths. Particularly, they wished to better understand the geology and basic science behind petroleum. As the 1860s wore on, many journals moved away from stories of the divine origins of petroleum in order to better understand its geological occurrence.

The impetus behind these articles could be explained as mere curiosity; however, the journals most frequently containing these articles were respected scientific publications, such as the *American Journal of Science and Arts* and the *Journal of the Franklin Institute (JFI)*. Often, a portion of the articles also sought to correlate the scientific aspect with the massive wealth being made in the oil business. This is not by chance. Explaining the origins of the oil greatly helped investment in the industry in two ways: by beginning to predict the supply's longevity, thereby helping people's confidence in investment; and by locating other oil regions throughout the nation.

An 1865 *Merchants'* article posed the million-dollar question, "if oil is found in one place, why not in any other? . . . if in one place only, or under certain conditions only, then in what place or under what conditions?" This article, geared toward the business reader, set out to discuss the geological origins of oil. It used scientific results to disprove the idea that increasing the number of wells would lower the pressure below, and thereby the amount of oil gotten. It concluded this discussion by trying to put into words the relationship between science and economic progress being played out on its own pages:

> While science pauses to ponder, to systematize, to classify, and to rationate, speculative enterprise, eager for the golden harvest, plunges into blind search, strikes out right and left, and is very sure to hit in the right spot sometimes. Science will follow at her slow measured tread, and only corroborate and substantiate the discoveries of the adventurer. This will prove so at least until facts have fertilized the waiting womb of science. Then will she bring forth an illustrious progeny, out of this, as out of other great subjects in natural history.[26]

In other words, don't wait for science to tell you to drill in certain places for oil. Readers should go out and decide for themselves; if they do so enough times, they will inevitably find something! Science and technology were clearly tools of industry in Petrolia.

As investment information became more in demand, advertisements for investment companies appeared more consistently than any other coverage in the popular media. These listed the stretches of land available, the company's capital, the prices of shares, the businessmen involved, and often descriptions of the land to be developed. *Merchants' Magazine* consistently ran more general advertisements with its listings of export data. One advertisement from 1863 (which carried no specific sponsor) acted as more of a pep talk for investors, relating that wells could be gotten, "with little labor, as the liquid wealth flows into his tanks as water from the ground. . . . But we notice that the most successful operators here are those who exhibit the greatest amount of energy and enterprise in obtaining and taking care of their oil."[27]

Articles in such journals often discussed the performance of specific companies and always contained detailed financial market information. The *New York Times* and *Merchants' Magazine*, for instance, frequently published trade and production reports of the young petroleum industry. The magnitude of the rise in the trade of petroleum could not help but be one of the greatest influences on potential investors. The consistent publication of these figures allowed investors to monitor their investments, while also illustrating to any investor the available potential.

With such varied coverage, it was not long until the first book on the region would reach the popular literature market. *The Oil Regions of Pennsylvania: Showing where Petroleum is found; How it is obtained, and at what cost. With hints for whom it may concern*, written by the *New York Times*'s William Wright, took the myths, realities, and economic possibilities of investment in petroleum and offered them to readers anywhere in the nation. Wright explained how one was to successfully invest in Petrolia and carefully discussed many of the scams being practiced, as well as a bit of the science involved in finding and sinking a good well. As a summation of this genre, the latter portion of the book teems with financial data.

Obviously, this book appealed to a popular interest in Petrolia and the universal eagerness to make money. This genre of coverage additionally sought to link one's providential right to make money with economic developments in Petrolia. For instance, one *New York Times* article concluded by offering its own biblical interpretation of Petrolia: "'to them that sit in darkness, light shall spring up,' that 'darkness shall be as the noonday,' and that in the home of the humblest man, 'at eventide it shall be light.'"[28] Aside from all private interpretation of prophecy, the article concluded that "there are simple and noble uses to be made of all the Master's gifts, if men have but the will to see them, and we hope that while Commerce claims new profits from Petroleum, and Science works out from its new results, Philanthropy will not forget to make it bear its share in the sweet services of charity."

This article presented a basic component of this genre of coverage, which was also a basic tenet of the American Gilded Age: money was to be acquired by any means necessary, but then a portion needed to be rationed back to the community. Petrolians appeared comfortable that some classes of Americans were intended to accrue more wealth than others, yet this boom also represented a universally accessible fortune. Interestingly, most occupants did not share the Gilded Age impulse to return some of the wealth to the community from which it had been gotten. For many oil developers, Petrolia was not a community at all—instead, it was a sacrificial landscape. Most philanthropic gestures were directed away from this ruined locale. This place, its investors seemed to recognize, would be abandoned. The location was its resource. Their charity should be given elsewhere.

Within the formation of any myth, danger is a major tool in stimulating the reader's curiosity. Historian Richard Slotkin's explanation of the frontier myth is surprisingly applicable when he writes that a basic American ideology "asserts that progress can proceed harmlessly," while literary mythology places conflict at the center of the story "and emphasizes the naturalness and inescapability of violence."[29] In this way mythic occurrences take on a duality based in the progress embodied by the myth and the realities necessary to accomplish that progress. For instance, an observer of writings of the mythic West never sees the impacted native culture and environment in the wake of the movement westward; there is only the successful farm and town on the prairie. Normally, the culture conceiving the myth resists ever realizing the violent transformation to focus exclusively on economic progress.

Petrolia did not consist of the dangers of raw wilderness and cultural conflict seen in the pioneers' movement westward. Instead, man-made objects, processes, and technologies composed the fearful potential of the oil industry. Petroleum's flammability combined with the American public's interest in reading about macabre detail to create an entire genre of popular literature based on the mythic dangers inherent in petroleum production.[30]

Each object, process, and technology bringing dangerous possibilities to this quiet, lightly populated valley drew fuel from economic aspirations—most basically, human greed. The duality of oil development could be seen on the face of the landscape: derricks and barreled oil signaling progress, but mud, excess oil, and the scalped forests illustrating the industry's impact. No one entered Petrolia without confronting this stark contrast. Each inhabitant and speculator needed to resolve this ethical quandary. The popular culture, on the other hand, made this quandary much more clear-cut for those who would never set foot into this valley by fusing the duality into a landscape.

Clearly, for local residents the landscape demonstrated the outcome of such new cultural mores through a great number of physical impacts, including the unparalleled occurrences of fire and flood.[31] As for outsiders, the popular press swept up tales of disaster in the Oil Creek valley and made them a major part of the myth of Petrolia. It was apparently assumed that if development proceeded at such a breakneck rate, the possibility of fire had to become a part of everyday life in the valley—even more so than in the typical nineteenth-century home.

No equivalent exists today to compare with the pervasive acceptance of the danger of fire in nineteenth-century society. In their study of fire in the United States, Margaret and Robert Hazen wrote, "The same form of energy that could build a city could destroy it; the heat that sustained life could, in excess, snuff it out. . . . A popular proverb put it more simply: 'Fire is a good servant, but a bad master.'"[32] During the nineteenth century open flames were still a welcome and necessary presence in every home. New dangers arose with the home use of highly flammable substances such as petroleum. At this time homes were tinderboxes constructed from wood.

This flammability, of course, would only intensify with oil's presence in the home. Fire and oil, petroleum, and kerosene were inseparably linked by a combination of high flammability and a shocking lack of care in handling the products. The experience of H. R. Rouse, described earlier in the chapter, demonstrated the popular culture's interest in such events. However, there often appears to have been a necessary purpose in including such details. Following the horrific account, the article concluded with a callous description of the amount of oil, wells, and total financial worth lost in the fire. These two tendencies represent the sensibilities of the nineteenth-century public aroused by a mixture of the morbid and the financial. It is these characteristics to which the popular press appealed with its accounts of oil fires, thereby also governing Petrolia's popular image.

In 1864 the front page of the *New York Times* announced the "most destructive conflagration that ever took place in the oil regions." A researcher on the topic would dutifully note this, only to realize that nearly every subsequent fire in the decade would also be given this label. Shortsighted journalism? Doubtful. Instead, the Hazens again attributed this to the nineteenth-century psyche: "Americans of the nineteenth century liked to measure their national progress according to . . . standards."[33] This was embodied by a constant effort to categorize and rank themselves and their culture. Their fires were no different. But to what end?

The accounts generally deviated little from one another. The only possible explanation for this coverage and its extension into an entire genre is that mythmaking was at work—the myth of Petrolia—and that it was meant to suggest other fiery images such as war and hell. It is important to note that many of the fire accounts reported in the *New York Times* were not written by their own correspondents. Instead, the stories were often picked up from papers in western Pennsylvania, a booster group who would absolutely benefit from making a myth of Petrolia.[34] Petrolia, the land of incomparable opportunity, was not meant to be a place to take the children!

One of the most exceptional portrayals occurred during 1866 when the national press covered a string of oil fires. Correspondents had been sent to Petrolia to get a firsthand look at the fires that had been burning through early February 1866. The account that they produced teems with a propagandistic air of grandeur. "An Artificial Summer—Vegetation Started by the Heat," exclaims one of the subheads. "Imagine," wrote the author, "a space perhaps forty feet square sending up a solid sheet of flame nearly sixty feet in height. It lights up the country for miles around, so that one can see to read a newspaper at a distance of four or five miles. The heat of the fire has started vegetation growing, and grass may be plucked there as green as that found in Summer time."[35] The correspondent offered the first portrayal of the people involved, when he described the irony of women and children being forced to scavenge for sources of fresh water in this region of financial splendor. "Oil might buy coffee and tea," he commented, "but could not make it so that urchins who had to [become] water-carriers appeared to think there might be too much of a good thing, even if it is oil!"

It became very common for these accounts to use instances of dramatic horror as a way of piquing reader interest before launching into what can only be called a publicity piece for the region. This article concluded with this telling sentence:

> Between the barren and inhospitable walls of the Oil Creek banks, in a country where the necessaries of life are all imported, where no infants prattle, and the smiling faces of women are rare, there is yet in progress a grand poem of humanity, dignified and adorned by all the alternation of success and disappointment, which develop and test the better qualities of the heart, and prove

men to be in truth "a little lower than the angels," even in the dust and confusion of life's stern and most forbidding conflicts.

The myth of Petrolia here becomes a grand poem of humanity. The horrors
remain as necessary as barrels in which to store the product. In good Protes-
tant order, material gain had to be accomplished at some price. In Petrolia
that price was damage to the community and the landscape.

Similar coverage continued through the mid-1860s. On March 23, 1866,
the *New York Times* reported the "greatest fire ever known in Oildorado." It
followed the standard reporting technique, concluding with the description
of one who was burned: "He was the most horrid sight ever witnessed, being nothing but a blackened, charred mass of flesh when extricated from the
devouring element, and he was in every way unrecognizable." Amazingly,
the standard was upgraded only a week later when the "most disastrous" fire
ever known in Oildorado took place, making the creek "a vast sheet of flame."
But, stop the presses, the standard was upgraded once again for a fire in Oil
City on May 26.[36]

With this profusion of fire coverage, the *New York Times* on June 1 used
the fire in Oil City as an opportunity for a general discussion of the "dangers
incident" to Petrolia. "The rates of insurance here are very high," wrote the
correspondent, "and the amounts of the policies are difficult to obtain."[37]
Specifically in the Oil City fire, the article stipulated that out of over $1 mil-
lion worth of damage, only $130,000 was covered by insurance. After an
itemized list of losses in the city, the article struck an emotional tone as it concluded, "It is a sad picture to look upon the ruins. The scene of a busy and
thriving place is now . . . one vast scene of desolation."

These scenes inspired only limited efforts at safety. Instead, the interest in
increasing revenue drove most technological advancement. The practice of
torpedoing wells discussed earlier introduced many dangers by forcing the
use and transportation of nitroglycerin through the region. One front-page
account from 1871 associated the horrific description seen in earlier accounts
of oil fires with the explosion of a wagon loaded with four hundred pounds
of nitroglycerin. "The shock was like that of an earthquake," reads the account, "extending for miles, and being felt throughout the surrounding
country." The description of the condition of the driver of the wagon is rem-
iniscent of earlier descriptions. His face was found, "almost entirely without
a skull. One eye was blown out, and the other was open, glaring and transfixed

in death. The mouth and nose and mustache were perfectly natural, and apparently he had not time even to put on an expression of alarm. All the remaining portions of the body were scattered so widely and torn into such fragments that it was difficult to tell which belonged to the man and which to the horse."[38] The account went on to give an equally gruesome description of the condition of the animal. In this fashion, added technology did not always mean increased safety. Often, it made the dangers even more extreme.

The lack of interest in enhancing safety ultimately illustrates that residents accepted fires and explosions as necessary by-products of the oil industry. Such occurrences formed a major portion of the myth of Petrolia, especially furthering the depiction of the region as a ceaseless inferno. In the end, people wanted to read of such detail, and so journalists continued to find ways of perpetuating Petrolia.

The years leading up to the twentieth century brought particularly drastic changes in American life. The altering interaction between humans and nature fed the myths of later generations. The oil saga's odd details, the complete unexpectedness of the discovery, the immediacy of its impact, and the awesome transition in the region's potential and wealth before and after oil made Petrolia more crucial than many other sites. The oil story made up a timeless myth—as engaging then as it is now.

For instance, the cultural attraction of this myth survived in feature films and television. The same myths of instant wealth have reached the twentieth-century American from films such as *Giant* and *Boom Town* and television programs such as *Dallas* and *The Beverly Hillbillies.* An entire generation of Americans knows the oil business through the experiences of the hillbilly family who discovered oil and instantaneously moved up the social ladder into high society. The Clampetts' tale is consistent with the earliest mythic components of life in Petrolia, as is evident even from the twanging country tune that introduced *The Beverly Hillbillies:*

> Well, listen to the story of a man named Jed, a poor mountaineer trying to keep his family fed.
> When one day he was shooting at some food, and up came a fountain of bubbling crude—OIL that is, black gold, Texas tea.
> Well, the first thing you know, ole Jed's a millionaire and kin-folks said, "Jed, move away from there. California's the place you ought to be."
> So they loaded up the truck and moved to Beverly.[39]

Jed's effortless discovery of oil adds to its mystique in the modern era; however, in figures such as J. R. Ewing, an organized, corporate side also became part of oil's contemporary mythology. The corporate side made an experience such as that of Jed Clampett all the more humorous.

From its earliest moments, the myth of Petrolia was one of wealth, desperation, greed, tenacity, and resourcefulness. Its landscape, however, was not tranquil like that of Beverly Hills, California. Petrolia was hell on Earth, teeming with mud, haphazard derricks, constant engine noise, shacks, and skiffs, all awash in flowing oil and blowing soot. The myth spoke to Americans of the 1860s who were desirous of waking their own boom, but also to those stunned by the stark contrast of the pastoral landscape with the industrial.

If the oil of Petrolia had gone the way of California's gold, this myth would be its predominant memory. However, the flow persisted and flooded out much of the myth with the industrial organization necessary to administer the greatest commodity of the modern era. The same individual involvement that had intensified the environmental impact of the early oil industry had also been the greatest component in the myth of Petrolia. Though it would quickly diminish with corporate control, it lingered enough in the popular imagination that twentieth-century Americans could enjoy Jed Clampett's escapades.

Early articles about the oil industry also demonstrate the vision of landscape impacted by unbounded industrial development. More importantly, they illustrate American culture's acceptance of the despoiled landscape as a necessary ancillary to economic progress. Trust in technological development became the guiding force as Petrolia defined itself as a place from the outside inward. People who knew nothing of and cared little about the long-term life of this region or its human community operated this place as an industrial zone of very limited meaning. H. R. Rouse briefly proved to be an exception to this rule, donating funds for roads and internal improvements. In his fiery death, the industry lost a bit of its young soul. The Oil Creek valley had begun its decline down a slippery slope of single-minded resource extraction.

William Wright's guidebook to the region is one of the few publications that sought to take Petrolians to task. He used the threat of floods as a representation of overall practices when he wrote, "The influx of capital has been so unprecedented, that some of its people may have imagined they can snap their fingers at the natural laws; but these will, in the end, assuredly vindicate themselves."[40] Wright not only spoke of the possibility of flooding;

more to the point, he noted an overall ethos with which the residents of the Oil Creek valley put their natural resources to work.

The meaning of or sense of a place comprises a deeply intimate and enduring characteristic of any locale. It presents local residents with the opportunity to make a lasting reputation for their community and to create the cultural mores they most respect. In Petrolia, the commodity usurped this right of definition, creating enduring exploitation at the hands of those who came from elsewhere to profit from the early industry.

The lack of concern about this locale is demonstrated by the British *Cornhill Magazine*, when it set out to cover Rouse's death. The writer acknowledges the great usefulness of the product being extracted from the wells of Pennsylvania, but also cites oil's problems. He describes oil as the substance that is "now imparting a taste of its quality to everything wearable, eatable, and drinkable, throughout whole sections of the Union, from the eggs and muffins you devour at breakfast, to the sheets in which you lie at night, the soap-and-water with which you wash your face, the towels with which you would gladly cleanse the oleaginous particles from your skin, and the railway carriage in which you vainly seek to escape your persecutor." Soon, he adds, an American and a Canadian "will be detected in society by his scent, as easily as a musk deer, or a civet cat." The author then describes a horrible oil fire he has read of taking place in Petrolia. The potential danger unites with the product's generally uncivilized qualities to support the writer's ultimate claim that the product should be banned from Great Britain until it has been suitably "purified." He writes that "a cargo of gunpowder . . . would be much less perilous."[41]

More important than the author's argument is what he ignored. The danger of shipping this material up the Thames or storing it in urban areas made up the rest of his lengthy argument. In its raw form, he left no doubt, the commodity was a danger to all and needed to be carefully administered and regulated. However, he never once showed concern over the safety of those in the Oil Creek valley, from whence this diabolical infestation derived. The danger and horror there were apparently necessary evils.

Similar to this British author, Americans grew uneasy when the mythic qualities of oil began to extend out of Petrolia and nearer their homes—the myth was becoming too near reality. An urban movement to regulate the storage of petroleum took shape in 1865. In that year, the *New York Times* reported the passage of an ordinance that forbade any oil to be stored within the city limits of Philadelphia. A month later, New York passed a bill that placed strict limitations on the amounts of oil that could be stored within the city limits. Nearby explosions and fires in areas without such ordinances, such as an 1866 petroleum conflagration in Jersey City, furthered such public sentiment.[42]

By 1872, petroleum's growing importance had placed even larger amounts in storage near urban areas. The writer of an article with the headline "Our Present Danger" disclosed that the lower part of New York City contained great stores of oil, "which are stored with an utter disregard of the inevitable consequences in the event of fire." Additionally, oil and grease were to be found throughout the city, and pharmacies using such substances were located on the streets just below City Hall.[43]

Letters to the editor followed this article's publication and told of dangerous oil houses surrounded by lumberyards. Such an image clearly reveals the degree of urban concern about oil's flammability. Of course, the city's dangers paled in comparison to those in the region from which this oil came. When contrasted with the total disregard for safety and efficiency in Petrolia, the urban reaction ultimately demonstrates that fires and other unpleasant or life-threatening issues were accepted as necessary by-products in the oil-producing regions.

Largely through its myth, Petrolia became a sacrificial region for Americans, one where conflagrations were expected—and the larger the better in terms of stirring interest in investment. The myth of Petrolia was an obvious reality for the blackened and scarred Venango County landscape. But viewed more broadly, conflagrations were symptomatic of the overall priorities with which the oil industry viewed the entire landscape. Americans were fully willing to write off this valley's future if it could provide them a steady supply of valuable crude.

The landscape of Petrolia stands as a landmark to this trust in technology's inevitable ability to lead to cultural progress. In the original *Petrolia*, the history of the region published in 1870, Andrew Cone and Walter R. Johns described this process that would move its way across the American landscape: "Though we had seen more inviting abiding places, we had never beheld one presenting a better chance to make one's first million in, and thus start fairly on the road to comfortable affluence."[44] Their vision, born out of the myth of Petrolia, would become a major portion of the ethics and values that led Americans on an industrial route toward a common "road to comfortable affluence."

Chapter Four

Oil Creek as Industrial Apparatus

The Oil Creek of this valley often reaches a span of 75 yards, but most frequently remains between 25 and 50 yards, and periodically dips below 20
yards. In this valley, the waterway seems to merit the designation of "river"
more than its given one of "creek." This breadth largely derives from the
creek's dual headwaters. The farthest north of these runs out of Clear Lake,
near Spartansburg, at an elevation of about 1,400 feet.

From this point, a narrow Oil Creek begins its winding 35- to 40-mile trip
to the Allegheny River at Oil City. After about ten miles, a second branch
meets this Oil Creek at Centerville from Lake Canadohta at about 1,300 feet.
Without giving up much to tributaries, Oil Creek then joins Thompson
Creek and Church Run before draining Pine Creek near Titusville.
Throughout this valley, many runs, or small creeks, merge with the down-

ward flow, the largest of which are Cherrytree and Cherry Runs. From this point, the river gains span and power as it drops almost directly south through a 14-mile chute of steep slopes of rock and soil to the Allegheny, which the creek meets at an elevation of about 1,000 feet.

Today, this valley makes up the quiet home of the Oil Creek State Park. What better to do with abandoned lands whose soul and sinew have been extracted than to make them into recreational sites? In the beginning and at the presumed end of its history, Oil Creek was and is a natural entity. But what of that middle portion of its history? What of those years when the entire landscape became overrun and literally remade by economic boom? During its boom years, each of the topographical and hydrological facts of this place remained important; in fact, they became critical factors in the industry's stability. However, entirely new considerations recast the value and importance of these details during the 1860s.

No individual knew the extent to which these ideas had changed the river better than the man who spent many hours looking at Oil Creek through his camera's viewfinder. John Mather grew intimate with this place as would few others. His views show how the rule of capture helped to shape this new meaning, yet they show another dominant force as well: clearly, Oil Creek served as the hub of commerce for the early oil industry.

As the industrial system developed around transportation outlets, Oil Creek became an essential portion of the oil network. Yet the connection between the young industry and the resource grew even more primary than this during the early years. Originally, speculators, scientists, and local observers unflinchingly believed that wells could only be sunk in the lowlands directly along the river.[1] Therefore, for the first three years the boom confined development to an even narrower locale than Venango County. The river's role in this economic system did not preclude it from continuing to function as a cultural hub for the local community. For reasons as basic as survival and ecology, the riverway remained a primary part of the local community.

The commodification of oil and the rule of capture's influence in turning this episode into an economic boom made certain that this valley would never be the same after Drake's discovery. A place was created that became mythical to the American public, and this intensified the region's use and exploitation. However, the myths and legends mattered little to those trying to live in the Oil Creek valley. The world around each of them, whether they were natives to the region or not, became unlike any other locale in the world.

These changes were the steps setting many communities up to fail—to be sacrificed. But for an entire locale to be sacrificed, it must first be dismantled in small parcels. This process can be observed by studying a single entity of Petrolia's landscape throughout the boom years. The photographic record of the valley makes such observation possible.

With so many colorful figures in early oil, Mather was in no way most noteworthy—in fact, he held few leases, none of which produced well. However, in hindsight, the recorder of this valley's wild ride might be the most important figure of the entire oil boom. The "Oil Creek Artist" (fig. 4.1) realized very early that he could make a living by recording the historic occurrences here, and so he spent the boom years traveling the valley by skiff and wagon. It is largely his work that allows us to see truly the landscape of Petrolia and to reconstruct the changes that occurred in it during the 1860s. His work and his own mode of transportation leave little doubt that Mather also very early on realized that Oil Creek functioned as the lifeblood of this industry and this place.

Mather's photos combine with other sources to reveal a valley in which every aspect of life and each facet of the natural environment became instrumentalized as part of the industrial process.[2] This change was possibly most obvious in the item most central to the culture, ecology, and economy of the region (before and after oil). Indeed, just as the juncture of the Tigris and the Euphrates served as the cradle of human civilization, the Mississippi as the cradle for the earliest human inhabitants of North America, and the Merrimack as the cradle of American industry, this stretch of seeming nothingness along Oil Creek represents the cradle of the greatest commodity of the modern era.

For European settlers, the haphazard terrain of this valley and region made travel by wagon and horse very difficult. Oil Creek offered the easiest access. Its often shallow waters made necessary the use of flatboats or skiffs powered by poling to the river bottom. Eventually, teams of horses and mules pulled skiffs of supplies, visitors, and settlers up and down the river.

In the late eighteenth and early nineteenth centuries, markets for lumber provided the impetus to fell the stands of forest throughout the region. During this period, the creek became the industry's mode of commercial transport for bringing logs downstream. The flow of the river was managed in a practice known as a "pond freshet." Canadohta Lake, which is more than one

Fig. 4.1. John A. Mather, Oil Creek artist (photographer)

mile long and one-half mile wide, served as the source for the freshets. The lake's average depth is thirty-five feet, but it ranges down to sixty feet in some parts. From Canadohta, Oil Creek drops directly south, past Titusville and then to Oil City.

Lumbermen cut timber from the surrounding hills and laid it in the lake or creek. During floodtide, then, the lumber floated down Oil Creek to the Allegheny and on to the markets in Pittsburgh and elsewhere. Originally named Washington Lake, this natural-spring-fed lake became home to a variety of water-powered grist and lumber mills as early as 1798. Soon residents learned that flood tides could be man-made by loading the lumber into the shallow creek and temporarily breaching the mill dams above the load. The tide would sweep downward and swell the creek so that the lumber could be carried down at any time of year. This tactic stabilized the region's lumber industry, and eventually tanning industries also took advantage of the special capabilities of the old-growth forest available here. For tanning, the skiffs would be loaded with the thick bark of species such as hickory, hemlock, and oak and floated downriver for use in treating leather.[3]

Access afforded by Oil Creek, as well as the reputation that its name verified, eventually guided Colonel Drake to attempt the first flowing well of oil along its banks. This movement marked the start of Oil Creek's remaking. The commodification grew so pervasive that during the oil boom, Washington Lake became known as Oil Creek Lake.[4] This purpose was twofold: first, following the success of the Drake Well, this name association enhanced the purchase of property and drilling rights on the land around the lake; second, the lake was destined to play an even more instrumental role in the transport of oil downstream.

Oil Creek served as a cog in the oil industry from its start, but the intensity of its involvement fluctuated. The earliest oil made its way out of the fields in long trains of wagons over the extremely bad roads of the Oil Creek valley or in skiffs pulled by mules and horses along the river. However, oil's abundance made undesirable the expense and time-consuming transport by horsepower (whether on water or land). Oilmen soon realized that they already had in existence a technology that would greatly enhance their ability to bring oil down Oil Creek and to market.

To the earliest speculators in oil, commodities and technologies were largely interchangeable. Such a perspective made it much more simple to instrumentalize the natural environment and make it part of the industrial process. The changed uses of Oil Creek would not have occurred without this perspective, which dominated local decisions about land use. It is through these decisions that one can observe the ideas of commodification and the rule of capture overtaking a culture. Armed with such logic, in the minds of valley inhabitants a skiff full of loose crude oil, which looked very much like a bathtub full of liquid tar, became the equivalent of a solid, round tree trunk that would float naturally atop Oil Creek. For the early speculators, using pond freshets to transport lumber was no different from using them to do so with crude petroleum. Soon, the dams along Washington/Oil Creek Lake were in full use to start pond freshets that would carry the skiffs of crude downstream.[5]

Although it led them to make only minimal adaptations, the speculators quickly learned that these two products were extremely different. Whereas lumber would simply ride out the tide unmarred, control of these loaded skiffs was impossible. The oil skiffs, filled with either barreled or loose crude, did not weather the freshet trip as well as the hardy timber. In addition, some of the mill owners along Oil Creek initially resisted closing their mills for the twelve to forty-eight hours necessary to prepare the loads and build up sufficient backwater.[6] The oil speculators banded together to hire a superintendent who would orchestrate the controlled release of water, and, even more importantly, appease the mill owners and keep them willing participants in the industry.

Divided up among many oilmen, the freshet cost proved minimal. The mill owners' charge for the use of the dams, the salary of the superintendent, and the services of two men to cut the dams made the cost of each pond freshet $100 to $400.[7] The price rose quickly in the early 1860s, owing to the expanded demand for lumber, which increased the amount of payment necessary to satisfy mill owners.[8] To defray the cost, agents traveled down Oil Creek just prior to a freshet and collected a toll on each barrel from every boatman. The price varied with the fluctuation in each mill owner's demand; however, there was no fluctuation in their demand to receive their payment prior to breaking the dams. The price in 1862 fluctuated around the sum of two cents per barrel; therefore shipments of no fewer than ten thousand barrels were required to meet expenses.

During the busy seasons of the 1860s, pond freshets were created twice a week. The superintendent set the date ahead of time so that skiffs could be towed to launches, overhauled, and loaded. Most boats held seven hundred

Fig. 4.2. Oil dippers, Miller Farm, 1863

to eight hundred barrels of crude or its equivalent in bulk if shippers wished to avoid the cost of barrels. If loaded with loose oil, the skiffs began leaking as soon as filling had begun, and leaking would continue until the oil was unloaded hours or days later. This was the logic present in much of the industry during the 1860s: the cost of lost oil remained less than the cost of barrels, and therefore skiffs could be loaded with loose oil. What happened to the lost crude or to the efficiency of the industry concerned no one; additionally, the loss truly cost little in the grand scheme of the industry.

Oil entered the river through skiff leakage and other lesser source points to create a level of pollution in Oil Creek formerly unknown to humans of the 1860s. Children using tin cups constantly moved along the river's edge in order to get close enough to skim the oil coating from its surface (fig. 4.2). Ordinarily, skimmers worked for personal profit—not for the safety of the people or for the good of the river. Of course, important safety reasons existed for skimming the oil from the regional thoroughfare: any time a fire would break out on land, the possibility existed that it would not only spread along the shoreline derricks but would also ignite the oil floating on the river itself.

No known method existed for putting out the fires on Oil Creek during the 1860s. They could only be viewed with horror as their flames rapidly spread downriver and even across to the opposite shoreline. A fire on top of Oil Creek could literally be observed moving downstream like a piece of debris making its way to Oil City. During the 1860s, Oil Creek was not a natural resource, river, waterway, or wetland area. Users of this era reconfigured Oil Creek as a mechanism in the industrial process of creating petroleum.

"Pull the shoats!" would be called out from up and down the river, meaning that all was ready for the freshet to begin. Normally, the superintendent and his men commenced cutting the dams around midnight. They would cut each dam in succession until they reached the last one, the Kingsland Dam, which lay three miles below Titusville.[9] When they cut this dam, the creek would rise as much as thirty inches above the highest outcropping of rock.

The rushing water brought with it a cool breeze as well as considerable excitement among residents.[10] The freshet signified progress, industrial advancement, economic prosperity, and, specific to Petrolia, great danger. When the river began its raging torrent, life and property up and down its banks were temporarily placed precariously on the cusp of disaster. In other locales that used freshets, the potential for damage may have been less because the riverscape was designed around such a function; however, in Petrolia, all the industry and development immediately bounded the river. Organizers granted little thought to avoiding disaster—however, its possibility excited many.

One or two boatmen staffed each of the hundred to two hundred flatboats. These men could only hurriedly try to control the skiff with a single pole as it descended the windy valley. This attempt, however, was like riding a bucking bronco—little guidance could actually be exerted over the skiff's path. Instead, these boatmen acted more as policemen over their boats' valuable loads. Without these monitors, it would be extremely difficult to track down a load if and when it reached its destination.

As these boats came tumbling down the river, the call "Pond freshet!" resounded through Oil City, the town at the meeting point of Oil Creek and the Allegheny River. All business momentarily stopped in town, the sleeping awoke, and the entire population proceeded to the riverside. They came to see the great flow of industry and economic progress, but also to be sure not to miss the disasters that often accompanied the freshet. Any freshet could produce one of the important historical moments that frequented the Oil Creek valley of this era. Every man, woman, and child who stood along the shoreline knew that if one boat went crosswise and stuck at any point, it would be crushed, as would others that followed. And that was often not the worst-case scenario.

Possibly no other industrial practice so clearly reveals the oilmen's nonchalance. There was simply no alternative for transporting the crude, and there was no competing source of oil in the world: *any* oil arriving in Oil City and beyond was a boon. One writer matter-of-factly described a basic problem during the freshet: "The boats are crushed against each other, and being generally built very light are easily broken, and if loaded with bulk oil, the contents are poured into the creek. If in barrels, the boat sinks, and barrels float off, and the owner rarely recovers them again."[11]

Additional costs prohibited heavier construction of skiffs. Hence, the broken splinters of lost boats and the commensurate abundance of crude pooling on the surface of Oil Creek littered every freshet. The consistent loss of so much crude allowed some of Oil City's more entrepreneurial inhabitants to improvise small dams from which they could collect the lost crude and sell it at their own profit.[12] After one particularly disastrous freshet, residents took out leases on land along the Allegheny at Oil Creek's mouth for collecting the excess oil from the water.[13]

Indeed, no matter how successful the freshet trip, skiffs lost a third of the loaded oil to leakage before the boats even started, and another third before the oil reached Pittsburgh.[14] Losses from collisions with other boats came on top of this already staggering amount. The Reverend Eaton's 1866 observation of the freshets applies to most practices in the valley: "This extemporized navigation is kept up and regulated by a kind of code of honor. Written laws and legal enactments have not yet learned of its existence."[15]

During the early 1860s, transient lessees and small holding companies based elsewhere served as stewards of the land along Oil Creek.[16] Pond freshets provide one example of the code of ethics that such land users practiced here during the 1860s. Such practices brought with them an unparalleled impact on the culture and natural environment, one that endured after the boom. At the center of this industrial scene, as it had also been earlier during the natural and agricultural scenes in this valley, flowed Oil Creek, now a full-blown apparatus of petroleum development.

An attitude of transience quickly dominated the environmental ethics of those in the Oil Creek valley and made possible scenes presented in many of Mather's photos of Oil Creek, such as one of the creek at low water around 1861 or 1862 (fig. 4.3). The practice of the pond freshet inevitably created a great deal of erosion in the valley, and, no doubt, Oil Creek widened over the years as a result. In addition, the placement of wells and equipment directly along the riverbed intensified erosion. Up and down either side of Oil Creek, wells were sunk and derricks constructed, creating a riverscape unlike any seen before or since.

The extraneous details within photos often reveal practices of the industry. For instance, in figure 4.3 we also see at least four or five derricks placed so near the river's edge that it is dangerous to the derricks' security as well as that of Oil Creek. Most startling, however, may be the aesthetic appearance of the water. While the primitive photographic techniques are partially responsible for the glossy film we see on the water's surface, this scene presents a dirty and murky composition that undeniably shows the effect of oil spilled into the river.

While oil seeped from casks and skiffs into the river, the proximity of the wells to the river caused additional large amounts of leakage into Oil Creek. The derrick structure usually incorporated a raised platform with a crude dam of thick timbers forming a square with the well at its center. These dams served many functions during drilling and pumping but did little to keep oil from seeping into the river. In addition, the crude technologies of the early years created oil leaks at every joint and turn. With the derricks near Oil Creek, all of this oil fed directly into the river. When development led away from the lowlands, the massive runoff still caused excess oil from nearly anywhere in the valley to run into Oil Creek.

When Oil Creek became the center for the developing industry, the initial developments demonstrated the boom mentality earlier referred to as the ethic of transience. Each of these sites began as a very localized lease, such as those seen in figures 4.4 and 4.5. As we can tell from the stability of its construction and design, the first site appears to have been part of a larger lease. The pump house in the center contained a steam engine, the force of which would then be channeled out to power two derricks. Particularly surprising in this photo is the use of paint on an industrial building, a frivolity rarely afforded such utilitarian structures. More often, these early scenes appeared like that in the next photo. Many trees remain standing, and a smaller steam

Fig. 4.3. Oil Creek at low water, 1861–1862

Fig. 4.4. Twin Wells, Blood Farm

building accompanies each derrick. In addition, a home is just up from the site, which suggests that this was a private lease on which the tenant could reside. At any rate, the impact of these localized sites remained great. Oil from each derrick seeped directly from the well into its surroundings, here (and in most cases during the early 1860s) into Oil Creek.

In figure 4.5, a barreling and loading dock abuts the derrick site. Normally, a spout fed from the larger storage tanks down to the barrels either near or in the skiffs, again greatly increasing the amount of oil spilled into the river. Often, oilmen placed much larger tanks dangerously near the edge of the river, which can be seen in a few later photographs. This 1867 scene shows a man sitting atop one of the scattered, abandoned partial dams that tried to control the high-water's flow. Over his shoulder, one sees a very large storage facility directly on the edge of Oil Creek.

These sites, however, seem tame compared with those industrial centers that sprang up on the Tarr Farm and on the three farms that would become

Fig. 4.5. Early creekside development for barreling and loading at water's edge

Petroleum Centre in the early 1860s. Such towns preceded the many temporary boomtowns that would appear later along Oil and Pithole Creeks. James Tarr and his family had farmed land along Oil Creek for many years. Partially due to Tarr's unique ownership of land on either side of the creek, the Tarr Farm rapidly became a major site of oil refining and packing. Additionally, the surrounding slopes were more gradual than at many other points along the creek, which made sinking wells more simple.

The Tarr Farm Petroleum Company administered the growth along this portion of the creek, including a refinery and a very large barreling works. What had once been a large agricultural tract soon became a haphazard conglomeration of derricks, storage tanks, temporary homes, and excess oil. An 1865 observer explained that this place was one of the valley's most successfully producing areas. "The owner of the Tarr Farm," he wrote, "in years past was a poor and uneducated man, who eked out a meagre livelihood by lumbering in addition to scratching the barren hill for a scanty crop. Poor as the surface crops may have been, the soil below has sent up products so rich that the lucky owner is now an exceedingly wealthy man."[17] The river that had formerly been at least partially a nuisance, because it bisected his holdings, now allowed Tarr to increase even further his possible earnings. Between 1857 and 1865, the Tarr Farm bought no additional land but leased neighboring land.[18] From this land, the senior Tarr generated enough wealth to retire soon to Meadville, Pennsylvania. He left his land to be overseen by those running the Tarr Farm Company.

The photographic record reveals that Oil Creek remained integral to the revised working of the Tarr Farm. Buildings and various objects litter the views, but it is not immediately obvious what exactly is being done in this place (fig. 4.6). In such a locale, the viewer must search the landscape for the mode of land use; when one does so, the contrast with agricultural pursuits becomes striking. In locations such as this, industrial change entirely altered the activity along the riverbanks.

Entire towns were thrown up along tracts such as this to increase proximity to the creek. None of the structures appears to be permanent in any way. The town, like the industry, is purely a product of boom. The houses, which are most often those structures possessing glass windows, are unpainted, simple, and usually one-room (which is suggested by the central chimney) cabins, which were built for less permanent habitation than those found in coal company towns.[19] The photo of Tarr Farm depicts additional

Fig. 4:6. Tarr Farm, 1865

farmhouses mixed in with similar temporary housing, derricks, and pump houses. The hotel most likely served as a boardinghouse for single men who had come to work the fields, suggesting that the ethic of transience infiltrated the traditional order of the region and its housing. In reality, who would have wished to live in this fine house amid the derricks, steam, and oil except for oilmen?

No better example of these boomtowns along the river exists than Petroleum Centre. Along Oil Creek's midpoint between Titusville and Oil City were three large oil boom "towns," McClintockville, Tarr Farm, and Petroleum Centre. The Reverend Eaton observed in 1866 that it was nearly impossible to ascertain where each ended, "so thickly strewn is the entire valley with a dense, active population."[20] Eaton estimated that the flatlands along this series of bends in Oil Creek were worth $250 million in 1866. With such incredible value placed on the natural landscape, how could any thought be given to values other than extracting as much oil as quickly as possible?

Fig. 4.7. Benninghoff and Stevenson Farms, Petroleum Centre, 1864

A conglomeration of land composed of the early, creekside Story, Mc-
Clintock, and Hays farms, the Centre proved exactly that. Just to the north
of the Tarr Farm, Petroleum Centre lay very near the creek's midpoint between Titusville and Oil City. In 1859, the five hundred-acre Story Farm was
sold to the Columbia Oil Company for $30,000 and commenced producing
oil in 1862. By the late 1860s, the site housed over sixty wells and three refineries. The Hays Farm, purchased by the Dalzell Oil Company in 1864,
possessed only scantily producing wells.[21] The G. W. McClintock Farm was
sold to the Central Petroleum Company in 1864 and housed the main portion of Petroleum Centre. Mather offers a clear view of the proximity of the
town's main thoroughfare to Oil Creek (fig. 4.7). One observer wrote of Petroleum Centre in 1865 that "here the wells are crowded as thickly as houses
in the most populous part of a city, dwelling and engine houses being mixed
up in such inextricable confusion that it is difficult to distinguish one from
the other without entering, and not always then."[22]

Fig. 4.8. Petroleum Centre, 1864

The town, which lay on the west side of the river, held the largest hotel along this portion of Oil Creek, as well as a series of smaller establishments not to be found in other boomtowns. Yet even in a developed town such as this, the industrial infrastructure remained omnipresent. In figure 4.8 we see another view of Petroleum Centre, complete with skiffs in the oil-coated water in 1864, revealing the elongated flatland that made this a suitable boomtown site.

We can follow a bit of Petroleum Centre's development by contrasting figures 4.8 and 4.9. The river appears heavy with debris and oil in each, but figure 4.8 also reveals another startling holdover from the valley's past: the two figures in the lower left corner, standing knee-deep in the polluted water, are dairy cows. The similar scene in figure 4.9 from 1868 shows additional structures and development, as well as the commensurate mud, oil, and debris, particularly in the area directly along Oil Creek. The still waters of this loading inlet appear entirely covered with a thick skin of oil, as is suggested by the one open pool in the top right portion of the river.

A lack of trees is evident in the background of many of Mather's photos. Oilmen deforested to clear the way for derricks and also to provide fuel wood

Fig. 4.9. Petroleum Centre, 1868

for the steam engines and industry of the valley. Most importantly, felling trees also cleared the way for the Oil Creek Railroad. The rail line ran directly along Oil Creek, providing the river with a new chapter in its history. Opened from Corry to Titusville in 1862 and then extended down Oil Creek to Petroleum Centre in 1866, the railroad greatly lessened the traffic on Oil Creek but did little to move occupants away from the creek because it largely ran south along the creek's floodplain.

Obviously, early oil speculators viewed Oil Creek as a resource. First and foremost, the waterway offered transportation of a variety of sorts (fig. 4.10). Additionally, the belief that oil could only be struck from the low-lying lands along the river made every square foot of the riverbank a valuable commodity. The flow of the river also determined the trade in crude, first in loading and docking sites (fig. 4.11) and then in the pond freshets, which together made Oil Creek an instrument in the industrial production of crude oil. However, the river's primacy to the developing industry may be best exhibited by the instances when a number of early speculators challenged riparian ownership rights and built the first offshore wells in Oil Creek.

Fig. 4.10. Blood Farm, 1860s

Figure 4.12 shows the miniaturized design found at each offshore well site. Here, the dam about the bottom of the derrick has been constructed to withstand water flow, as well as to contain the oil from the well. Often, the dams were constructed in the shape of a diamond, with the points facing up- and downriver.[23] Many of the steamhouses employed steel extensions that reached across to wells on the other side of the creek; however, this well appears to have been run by a steam engine contained within the shack structure. The derrick appears to rise directly from the roof of the shack, providing a wonderful example of the improvisation of much of the early technology in the industry.

One must wonder about the consequences for such a well during a freshet. One observer of an offshore well in 1865 seemed to take the phenomenon in stride, as one among many oddities in the valley. He observed only that in the middle of the river was an old well where once there was known to be a bubbling spring of oil, but it had not "paid out well."[24]

The increasing instrumentalization of Oil Creek carried a severe penalty. The values and ethics of Oil Creek valley residents exhibit themselves in people's inability to learn from the industry's severe toll upon Oil Creek. Despite horrifying fires upon the water, for instance, valley residents made few changes to lessen the amount of excess crude flowing into the river. Residents again and again viewed the river as an extension of the industry and commodified it as such. Any damage to it became a necessary ancillary to the industry being carried out. Repeatedly, however, the river brought lessons of the valley residents' affinity for believing that they existed outside natural laws and limitations.

After being instrumentalized, Oil Creek could cause significant human disaster by simply following its natural, seasonal patterns. During many

Fig. 4.11. Creekside loading site

Fig. 4.12. Offshore derrick

high-water seasons, Oil Creek brought damaging floods throughout the valley. These were at least made more damaging—if not altogether caused—by the erosion and runoff created by the creekside development along each side of the fourteen-mile stretch of this valley. For instance, in December 1862, while the loaded fleet of boats lay at anchor, an ice gorge broke loose and came down Oil Creek, crashing against 350 boats filled with sixty thousand barrels of oil and destroying nearly half the oil, valued at $350,000.[25]

In May of 1863, the sly river once again carried near disaster to the surrounding community; this time a lantern set fire to a bulk-loaded boat of oil lying at anchor among four hundred others below Oil City. The flames quickly spread to at least fifty-two other boats, many of which were then cut loose to spread their flames across the eddy and on to the oily land on either side. The entire area was soon on fire, with burning barges extending the flames down the Allegheny River as far as Franklin.[26] In typically shortsighted hyperbole, the editor of the *Spectator* dubbed the occurrence the worst disaster in the history of the oil regions.

Disasters could take other forms as well. The greatest of the many pond freshet accidents occurred in 1864 and is recorded in a stunning series of photographs. In this photo we see hundreds of flatboats piled up near the railroad bridge in Oil City (fig. 4.13). As the modern observer views these scenes, it seems unthinkable that the freshet practice was not altogether abandoned following the disaster. The facts that freshets continued and that they would only be abandoned when a more profitable mode of transportation had been devised demonstrate that Oil Creek was managed exclusively as a device in the overall industrial process.

The presentation of Oil Creek to observers outside the valley took a variety of forms. In 1865, J. H. A. Bone traveled to the region and then quickly published his account as *Petroleum and Petroleum Wells*. He would first see Oil Creek from the railroad tracks that wind through the region and then directly along the edge of the creek.

> Presently the derricks increase; they close up their ranks, and soon stand in unbroken line along the left bank of the stream, throwing frequent skirmishers across to the right bank, effecting lodgment at the foot of the precipitous cliffs, where there is barely room to stand, and even threatening the railroad track which winds higher up. . . . The river is dark, and a scum of oil glistens

Fig. 4.13. Pond freshet disaster, Oil City, May 1864

on its surface. Here and there a small board-shanty, connected by slender pipes with tanks at a little distance, marks the existence of a refinery—for all the processes connected with oil, from its extraction from the rock until it is ready for consumption, are carried on in the vicinity of the wells.

In this version of the Petrolia myth, the author's use of military imagery suggests the derricks as soldiers blazing the trail through this valley toward economic progress. Concerning travel through the region, Bone called passage by flatboat an "abomination" and offered that travel by foot was the best alternative. Although, "there is a choice of paths in going down or up the Creek, the difference being that each is muddier than the other, and that you are certain to select the muddiest." Due to the thick mud, Bone added, there is always the possibility that one's next step may "descend deep enough [that] you may strike oil."[27]

While Bone's presentation of Oil Creek held back none of the base details, some others made an obvious effort to present the creek as a scenic commercial thoroughfare. Oil Creek's depiction in lithographs grows out of that medium's widespread effort to offer models of communities and landscapes that were similar nationwide. In these lithographs, one sees that including Oil Creek in this typology outweighed any commitment to realism. The scenes differ significantly from the reality of the written and photographic accounts. While freshets were the predominant method of transporting oil on the river, *Harper's* published an illustration of teamsters serenely carrying a particularly orderly load of barreled crude downriver and a cargo of passengers, respectively (fig. 4.14).[28]

These types of lithographs were a major portion of the effort to market the region and the creek to potential investors. F. W. Beers's 1865 *Atlas of the Oil Regions* was the single most frequently used recruitment tool for speculators.[29] The carefully constructed maps reveal the region as a commodity available for oil speculation. Leasing lots, each bearing an identification number for use by potential investors, line the river's edges. Following maps and listings for assorted townships, a sectional plan offers Oil Creek as the valley's unifying factor. The entire organization of the region rejects the impermanent towns and grasps the stability of winding Oil Creek.

A work such as Beers's *Atlas* functioned like Dr. Silliman's earlier report on the commercial potential of the product: quantifying the scene to the American observer who had long heard of the region from the popular press.

Fig. 4.14. Transporting visitors on Oil Creek, *Harper's*, lithographic image

In one map, from Oil City to the north lease tracts are disturbed by few land-holders who have not parceled out their land. The numerical designations signify lot numbers to which investors can refer when talking with leasing agents and company representatives.

The *Atlas* made it possible for the distant observer to be directly involved with the place by associating specific sites with depictions found in guide-books such as those written by Cone and Johns, Bone, and Eaton. In such a fashion, the *Atlas* completed the process of making Oil Creek one with the commodity petroleum. Whereas the creek had previously been viewed as an extension of the industrial process, it now additionally functioned as the basic structuring agent for long-distance financial speculation.

Petroleum's rise from being merely a point of reference for locating this region to its essence and reason for being mirrored the same transition in the perception of Oil Creek. This emergence began with Drake's well and then continued because of the industry's belief that wells could only be sunk along the lowlands bordering the river. Finally, the river's primary role as a route of commerce left it little meaning beyond its function within the industry. It is an interesting irony that this exploitation and the horrendous impact that it exerted upon Oil Creek have, in the end, largely enabled the river to return to a tranquil existence that it had not known since the first inspection by Europeans in the late 1700s. Oil Creek has earned its repose.

*In brief, we have here in the Oil Region
an Utopia, almost, if one is disposed to
seek only for that which is good.*
—Andrew Cone and Walter Johns, *Petrolia*

*The trade is flourishing as no trade ever
did before. . . . [Petroleum] is attracting
more attention and interest than all other
branches of Trade or commerce. . . . the
very nation, torn and bleeding and
suffering, looks to it as the great resource
from which it is to draw its recuperative
energy now that the war has drawn to a
close. Its appearance is that of a vigorous
young tree, luxuriously covered with
leaves and blossoms, and full of promise
for the future.*
—Rev. S. J. M. Eaton, *Petroleum*

Chapter Five

"What Nature Intended This Place Should Be"

In April 1864, the 450-acre Blood Farm, a large tract of land bordering Oil
Creek on each side, sold. It was a fine farm, half of it fit for immediate culti-
vation and possessing a large dwelling house, a barn, and many smaller buildings. At a time of furious land sales, this transaction seems simply like any
other; but this sale sealed the fate of the valley's present and future. Blood
Farm brought a price of $550,000, a staggering sum even in Petrolian terms.
But even this did not indicate its true importance to the entire valley. In fact,
the *Venango Spectator* supported the purchase price by writing that "of course
we need not tell the reader that the agricultural value of the farm is nothing
in comparison with this investment over half a million. The value is in the
oil already producing and in prospect."[1]

In retrospect, the valley's future became more and more obvious during the early 1860s, yet some residents maintained their farms, while others altogether resisted the money being offered to them by oil developers. In the early 1860s a hodgepodge of owners and even types of land use could still be found along the lower Oil Creek valley (map 5.1). But all this would soon change. The *Spectator*'s correspondent described the Blood Farm land as "the only piece of oil territory on the creek not [yet] in the hands of speculators." With this sale in 1864, the entire riparian lands along Oil Creek were owned by new landowners who were prospecting for oil.

With Oil Creek securely and completely commodified, the entire valley became a mechanized, industrial locale designed only around the idea of getting its oil out to the world. The vast majority of these individuals came from far-off towns and cities. They knew nothing of the valley except its ability to produce this rare commodity. And that is just the way the industry wanted it.

In 1864 the *Spectator* estimated that Blood Farm was already within the top five most productive sites in the oil region, without yet having endured major development. In one year of intensive development, the author estimated, the farm would bring in $720,000, making the purchase price seem low. As the writer stated, "It is a big thing and a sure thing." The author concluded by revealing his reason for detailing this transaction: to prove that half a million could be put into a single farm and be "the most profitable investment of the day." By the next year, the Blood Farm land along Oil Creek had been cordoned off into forty-five large leases and subleased many times over.[2] The general area, however, was still called Blood Farm, as it is yet today.

Just as they did to Oil Creek and Blood Farm, economic and ideological changes altered the lives of everyone in the valley. This analysis stresses the physical landscape and the use and ethics that define it; however, there is also a human element crucial to the landscape. After all, nature does not create a place such as Petrolia; culture does. Economic considerations drove the culture that created the landscape of Petrolia—about that there is no doubt. Much like parts of Wyoming with ranching towns, Maine with fishing towns, or Illinois with farming towns, these one-dimensional communities depended on extraction.

As this society and culture took shape around the acquisition and processing of crude oil, many residents and onlookers came to believe that this region was a high point of human civilization. Where the formerly unused

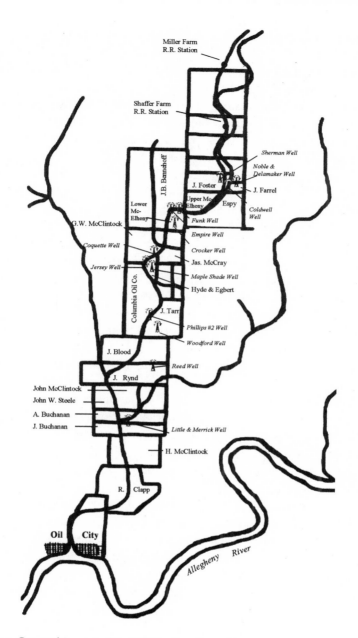

Map 5.1. Ownership map of the Oil Creek valley, 1860–1868

Sherman, 1,500 B/D; Noble and Delmater 3,000 B/D; Caldwell, 1,500 B/D; Funk 3,000 B/D; Empire, 1,500 B/D; Crocker 3,000 B/D; Coquette, 1,500 B/D; Jersey 3,000 B/D; Maple shade, 1,500 B/D; Phillips #2 3,000 B/D; Woodford, 1,500 B/D; Reed 3,000 B/D; Little & Merrick, 1,500 B/D.

or wasted became useful and valuable, Petrolia, some said, became "all that nature had intended." Human resourcefulness, and specifically its application through capitalism, created such economic progress, and many thought it would then lead toward social and cultural progress. Through the grime and the waste crude, many Americans discerned such a future for Petrolia during the 1860s.

These aspirations for permanence would ultimately hold little weight in regional development. However, its impermanence does not mean that a culture failed to develop in Petrolia. Instead, the effort to allow this region to attain its "natural" potential directly and indirectly influenced the culture of those participating in the new industry. As boomers came from throughout the state and nation, shifts in ethnic diversity, age, and gender altered the composition of the population. The fashion in which these people existed, made their living, and resided near massive industry completely changed social patterns throughout the region. These changes dominated some towns, such as Oil City, which became a glorified boomtown. Others, such as Franklin and Titusville, worked actively to limit the extent to which their communities became an industrial work zone. Even so, no community in the region was left untouched.

One of the oil industry's most legendary tales illustrates the difficult coexistence necessary in oil communities. Rial and Son, a Franklin company, drilled its Number 9 well on Point Hill, a small mountain near Franklin, and found a showing of oil at 490 feet—200 feet earlier than normal. In addition to a streak of oil, the pool contained an additional substance. Suspecting what it might be, the drillers sent for Philip Grossman, the local brewer. He reported that it wasn't just beer, "it's mein own make. You are pumping mein beer vault dry!" It turns out that Grossman used one of the caves at the foot of Point Hill to house his vault for aging his beer. The driller had pierced the large keg in the vault and was now pumping it out like oil![3]

While the petroleum industry defined itself and its overall processes around the boomtown or work camp, the efforts and needs of a few communities to manage the boom while keeping their community—and beer supply—intact compose one of the most unusual stories of the early boom. The spearing of the underground beer vault is an example of incidents in which the oil boom infringed on ordinary life in existing communities; most of the events weathered by these more established towns, however, were of greater consequence than a loss of beer.

To reside in this valley during its boom required that one abandon certain expectations of comfort and civility, enabling him or her to place the generation of oil and individual income as the primary motive for his or her actions. If done on a massive scale, this way of life changes the meaning of an entire place. Even though an oil field was called Blood Farm, for instance, it had changed in every aspect from its agricultural predecessor. During the 1860s a local culture supported these alterations and fully accepted wholesale changes in life, technology, and economics as a matter of course. Just as it created gushers, great wealth, and an industrial infrastructure, the early years of oil also inevitably made a culture out of the chaotic land rush that became boom.

The relationship between culture and technology is always a close one, with each intermittently influencing the other, but with culture ultimately initiating the drive for innovation. In a situation such as Petrolia's, technology becomes the defining point of the entire culture, creating new values and everyday patterns of life.[4] The motivating factors behind people's actions and decisions often reveal a dramatic change. As people strive to turn cultural values into physical realities, a community's landscape takes shape.[5] This cultural artifact then remains for inspection and analysis.

Geographers and nature writers use a wonderfully vague and sweeping notion to encompass these cultural revelations: a region's "sense of place." This sense is different with each geographical location discussed, and would certainly be altered by an occurrence such as massive industrial change. Historian Donald Worster describes the sense of place as "a complex adaptiveness in which the self reflects the community and the community reflects the natural system, and out of these interdependencies emerges a peculiar cultural ecology."[6] This definition sounds too scientific, too abstract, until one accepts the human element as just one element in the cultural ecology of a locale—but one that dominates nearly any ecosystem in which it resides.

Recast through great technological change, Petrolia succumbed to domination. Even though areas like Tarr Farm and Blood Farm maintained their previous names, the activities going on there in no way resembled an agricultural establishment (fig. 5.1). Basic alterations in ways of life and individual philosophies or ideas formed quickly, and like a pinion each turned on the production of petroleum. The census bears out this sudden change: statistics did not reflect the production for 1860, but by 1870 the fifth largest

Fig. 5.1. Cherry Run, 1867

mining industry in the nation had appeared from nowhere.[7] Technology carried out this growth: more than half of the nation's entire supply of steam engines powered the new industry, creating more horsepower than any other mining industry except anthracite coal. It involved many people at a variety of levels and provided each with new opportunity. There were more establishments related to the industry, 2,314, than to any other mining industry. Yet none of these were the statistics that really moved people to come to Petrolia or to invest their dollars in its young industry.

Most impressive, in only ten years the Oil Creek valley had put a value on a formerly valueless substance; and not just any value. By 1870 petroleum represented the third most valuable mined resource for the nation. Incredibly, petroleum followed only anthracite and bituminous coal with a sales value of $19.3 million. However, while each of these coal industries involved 40 to 50 thousand workers, petroleum involved only one-tenth that number, 4,488. With a less sharply defined hierarchy in place and fewer overall workers, boomers could acquire a large chunk of the $19.3 million purse. Such facts compelled many individuals to rush to the sites of the new industry. And if they were to do so, there was really only one place in the world to go.

Eighteen million dollars worth of the world's petroleum production—fully 94 percent of the overall total—came from Pennsylvania alone. In that state, 4,070 workers staffed the industry and its 2,148 establishments, each of which accounted for over 90 percent of national totals.[8] The industry involved risk anywhere, but Pennsylvania possessed the most established infrastructure with which to manage its supply. It was really the only place to go, and so the boom came (graph 5.1).

Within Pennsylvania, seven counties reported petroleum production (tables 5.1–5.3). Of the state's 2,148 establishments, 1,668 could be found in Venango County, home to Oil Creek and its famous valley of oil.[9] This one county contained 78 percent of the state's oil industry and 72 percent of that of the entire nation. In Venango County, 3,085 male workers labored in the petroleum industry, along with 1,662 steam engines able to generate 15,113 horsepower. In Venango County alone, nearly $6 million of capital was invested in the young industry. Finally, and most telling: in an industry whose product grossed $19 million nationally, Venango County produced 74 percent of this total, with annual production of more than $14 million!

As one might expect from such numbers, Venango County experienced one of the largest population explosions in the state during the 1860s. From 1790 to 1860, the Venango County population had never increased by more than 8,000 persons in a decade. Between 1860 and 1870 Venango County's popu-

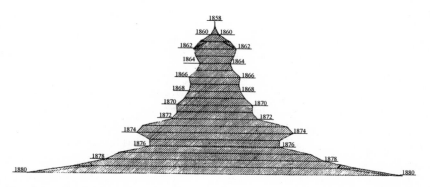

Graph 5.1. Production, 1859–1880. From Bureau of the Census, *Census 1880*

Table 5.1. Oil Production, 1859–1873

Year	No. Barrels
1859	1,000
1860	500,000
1861	2,113,000
1862	2,056,000
1863	2,611,309
1864	2,116,109
1865	2,497,700
1866	3,507,700
1867	3,347,300
1868	3,715,700
1869	4,215,100
1870	5,650,000
1871	5,302,710
1872	5,965,635
1873	9,882,010

Source: Adapted from Bureau of the Census, *Census 1880*, Vol. 10, "Petroleum," 149.

lation grew from 25,000 to 48,000—nearly doubling.[10] Such incredible growth drove away many previous residents as the county shifted from agricultural pursuits to those of the oil industry. The surrounding counties had similar growth rates, including Forest County's remarkable growth from 898 occupants to 4,010—nearly 350 percent.

With such shifts in total population, the complexion of the region's ethnicity grew more diverse. A remote enclave of German farmers prior to the discovery, this place became a melting pot to rival many coal company towns.[11] While the total population increased by 68 percent, the foreign-born segment increased by 432 percent. Events such as the oil boom offered a wide variety of opportunities, regardless of ethnicity and economic class.

This ethnic diversity suggests that Petrolia possessed a role in the American movement westward.[12] Acting as an usher for the movement further westward of many second-generation immigrants, Petrolia and its available jobs attracted many recent immigrants.[13] These individuals likely continued westward after earning income in the oil boom. Of the foreign-born popu-

Table 5.2. Selected Producing Divisions and
Districts, 1859–1873

Year	Oil Creek Division	Pithole District	Lower Allegheny
1859	2,000		
1860	500,000		
1861	2,113,000		
1862	2,056,000		
1863	2,011,300		
1864	2,116,100		
1865	1,585,200	912,500	45,000
1866	2,502,700	1,005,000	918,044
1867	2,396,300	954,000	1,001,458
1868	3,072,617	547,500	1,058,000
1869	3,702,500	305,000	4,402,563
1870	3,030,526	173,565	5,100,265
1871	2,040,365	182,054	4,712,702
1872	1,528,865	145,065	4,755,623
1873	1,094,389	119,964	5,431,072

Source: Bureau of the Census, *Census 1880*, vol. 10, "Petroleum," 149.

Table 5.3. Oil Field Development, 1859–1880

County	Acres Developed
Venango	32,000
Crawford	6,400
Forest	1,920
Warren	6,720
Armstrong	5,120
Clarion	19,200
Butler	27,520
McKean	50,000

Source: Adapted from Bureau of the Census, *Census 1880*, vol.
10, "Petroleum," 150.

lation in Venango County in 1870, the countries of origin were, in order, Ireland, British America, Germany, England, and Wales; fewer than one hundred individuals came from a sprinkling of other European countries.[14] The Irish population contributed most to the evolving ethnic character of the boom. While rarely becoming rich enough to own land, the Irish dominated the teamster and mechanical trades.

The uprooted were likely to move again. By 1880, the foreign-born population had dropped by 30 percent.[15] Inter-state migration followed a similar pattern. Beyond those born in Pennsylvania, the largest segment of the 1870 population had come from the state of New York; Ohio followed, a distant second. By 1880, half of each group had left Petrolia. Interestingly, each state followed Pennsylvania's path, becoming significant players in oil production and refining. Most likely, many of these individuals returned to their home states to disseminate the technology that they had learned in Petrolia.

The boom mentality loosened many mores of midcentury American life, yet racial ideas showed little progress. The African American population increased significantly but remained a small portion of the overall population. In Venango County, the number of blacks rose from 68 in 1860 to 433 in 1870.[16] With a great need for physical labor and service positions, the oil boom offered blacks good—but never equal—opportunities to make a living. Boomtowns often showed higher concentrations of blacks, which is most likely due to the need for quick labor. There remains no evidence, however, that the loosened culture and society of the boomtown treated blacks more equally.

While diversity became a hallmark of the workforce's ethnic makeup, gender and age demographics reveal patterns of homogeneity. Basically, oil attracted middle-aged males. Before the boom, the European American (white) population was divided almost evenly among the ages from 1 year to 50; however, the largest segment by far was younger than 40. By 1870 the gender balance had been skewed by 3,500 extra males. Additionally, the number of males aged 18 to 45 nearly tripled.[17] The boom demanded a staggering increase in the number of men to operate the machines and provide labor. While earlier histories have surmised that these men left their families behind, individuals counted by the census reveal a different pattern.

The population of women in Petrolia grew at nearly the same rate as that of men.[18] Overall, the European American population of women in Venango County increased by 95 percent, which is just below the male increase of

106 percent. These women were not involved in the oil business, which was entirely male-dominated on all levels. Most likely, these women, if members of the workforce, were involved in the service industries springing out of the boom, such as hotels, restaurants, and stores.[19] These women made up a significant portion of the workforce in Franklin, Titusville, and Oil City, each of which functioned as jumping-off points for speculators and as headquarters for trade. In boomtowns, on the other hand, males dominated the local population due to the communities' focus on industrial extraction.

The society shaped by these people during the 1860s little resembled that which had existed prior to oil. The melting pot of diversity, the new hierarchies of wage and labor, and the constant change of a population in motion defined a new social system in the Oil Creek valley. During the 1860s, this society formed a unique culture as well as the communities in which they would reside.

The oil boom defined towns in the region for years to come but did not necessarily follow the same process in each. Some resisted change as they could; others boomed like July Fourth fireworks, then faded to nothing. Oil City led yet another system of town development, which fell somewhere between these two. As boomtowns lost thousands in population during the later 1860s, towns such as Titusville, Franklin, and Oil City expanded significantly. Today, Oil City remains (as do the other two); yet, as its name suggests, it has been completely dominated by the industry from its first moments of existence.

The aptly named Oil City is intriguing because, like a boomtown, it was and is entirely a product of the oil industry. Prior to the commodification of oil, Oil City housed only limited permanent settlement. From nonexistence in 1860, Oil City gathered 2,276 occupants by 1870. During the next decade, however, when many other oil towns were collapsing, Oil City's population jumped by 221 percent, a rate of growth rivaling that of many boomtowns in the 1860s.[20] In essence, the population increase of Oil City indicated overall patterns within the industry, particularly the growing need for business administration as the industry grew and incorporated.

From a speculative industry involving individuals and small operators who would base themselves in work camps or boomtowns, by the early 1870s the industry had become a standardized, corporate endeavor. The business required a few large towns or cities from which the undertakings could be overseen and financed. This made it possible for many of the important corporate headquarters to remain in Oil City, even when other regions of the world became the larger producers of oil. An important example is the Oil City Oil Exchange, which was headquarters for the regional oil decisions of John D. Rockefeller and Standard Oil.

Even so, Oil City's development was more like that of a boomtown than of a longstanding corporate city. Most important, none of Oil City's founders were local. In early 1860 only 25 residents occupied the loosely formed village at the mouth of Oil Creek. That year, the Michigan Rock Oil Company acquired a large portion of modern-day Oil City and laid it out in lots.[21] Word spread, and speculators flocked to the town, often using it as the jumping-off point into the "Valley of Oil." In 1864 the Laytonia Town Oil Company set up the southern portion of the city.

Yet another group of investors established the northern portion of the city in the same year but did so under the name Imperial City. The two towns were then consolidated by a court order in 1866 to form a borough known as Venango City, with a population of roughly 1,500. Finally, the *Titusville Courier* reported in 1871 that "Mr. Oil City [had been] . . . married . . . to Miss Venango City."[22] In this gendering the people followed the patriarchal practice of the wife's adopting the husband's name: the union formed Oil City.

The first planners viewed Oil City as a river port. Surrounded by mountains and dense forest, the main oil-producing parts of the valley were very difficult to reach. Titusville, to the north, was equally inaccessible. By and large, Oil City became the typical approach to the oil regions before travelers ventured north by river. Hotels, banks, saloons, and stores filled the basic needs of oilmen throughout the region. Oil City quickly became one of the major bedroom communities for the Oil Creek valley.

Yet boosters saw even more potential for this locale. One 1866 booster publication explains that Oil City's facility for manufacturing "cannot be excelled. . . . All those engaged in the construction of the vast amount of engines, machinery, tools, barrels, and in fact, all the mechanical pursuits applicable to the petroleum business, might here find lucrative employment." In addition, the publication describes the beautiful vistas overlooking Oil Creek and suggests that these are similar to "the most picturesque portions of the scenery of the Hudson."[23]

Industries to support the oil boom also developed rapidly in the 1860s. No employment category in the 1860 census reflects involvement with petroleum.[24] In 1870, census takers added two categories entirely composed of those working in petroleum. Nationally, 1,747 persons were engaged as oil refinery operatives and 3,803 as oil well operators and laborers.[25] Of these, 459 and 3,567, respectively, were located in Pennsylvania. In terms of well operators and laborers, this is an astounding 94 percent of the national workforce.[26]

The new industry demanded machines and their maintenance. For most laborers in Oil City, the boom brought technology and the machine directly into their lives for the first time. No calling existed for mechanical trades prior to 1860, but by 1870 the manufacture of machinery involved 23 establishments and employed 164 workers. Whether in these businesses or others in the region, the workers' entire universe had been altered by their proximity to technology. Ideas of time, labor, scale, and waste would be restructured under a new rubric.

Mechanization grew out of pockets such as Petrolia but influenced the entire nation between 1860 and 1870. During this period, the national value of materials generated through mechanization increased by 91 percent. Pennsylvania passed New York to become the most industrialized state in the United States as the value of its products related to machinery increased by 141 percent.[27] The machine would define the American future, and potentially no region was so intimately involved with this process as Petrolia.

The machines and their oil boom brought related opportunities for economic development outside the oil business. The economic possibilities changed greatly for everyone living in Oil City during the 1860s, whether they were in the oil business or not. Cooperage, for instance, was directly related to the industry because the oil was stored and often transported in wooden casks of various sizes. In 1860, Venango County possessed one cooperage, which carried $1,000 of investment and employed 5 males.[28] In 1870, this single cooperage had been transformed into an industry of its own with 22 establishments and 148 employees.[29] A few years later, the Oil City Barrel Works alone employed 250 men and boys and manufactured 1,000 barrels a day.[30] Other growth businesses included brewers of malt liquors and wagon makers, both of which were spurred by the oil boom.

Such a dramatic shift in the type of work in which most of the society participated sent waves of change through the regional culture as a whole, but this should not suggest that Oil City made no effort to be a permanent community. The development of churches, for instance, is one of the most obvious examples of residents' interest in fostering a longstanding community. Religion and places of worship help to compose the core of any human community, but they are particularly crucial in frontier or developing regions, in which religion composes a basic structuring mechanism. Oil City possessed four churches in 1869, whereas it had none in 1860.[31] Throughout Petrolia, churches developed rapidly during the 1860s as communities sought to offer transients a way to structure their lives.[32] Presbyterian and Methodist churches made up the religious foundation of the oil regions.[33]

In an era when religion and church maintained a firm hold on humans, the rise in population necessitated additional churches. Irish immigrants quickly helped Roman Catholicism become the region's most rapidly growing religion. From a single Catholic church in 1860 that accommodated 300, the year 1870 saw 12 such churches in Venango County, which could accommodate 5,000 churchgoers. Similarly, it is likely that the infusion of Germans fueled the addition of four Lutheran churches that increased their accommodations from 650 to 2,200. Oil City's four churches reflected its European but ethnically diverse population: Presbyterian, Methodist Episcopal, United Presbyterian, and Catholic.

Even amid such hallmarks of community and society, there could be no doubt that all of Oil City and Venango County's economic eggs were in a single basket—oil. There would have been no other economic developments in Oil City without a steady flow of crude. The offices of various oil companies and of speculators and shippers located themselves along Main Street in Oil City's Third Ward in the late 1860s. Within the humble hub for the world's oil supply, a section of Centre Street served as the major trading area until speculators took it indoors in 1869.[34] Such trading made the local economy and society function.

With economic development as the industry's overriding priority, considerations of safety lapsed. In Oil City, for instance, residents lived every moment with the realization that it could be their last. The most pressing vehicle for this sensibility was the threat of fire and floods, which haunted every occupant of Oil City and cities like it. As Charles A. Seely wrote in the *Scientific American*, "The blase gentleman who looks down the crater of Vesuvius and finds nothing in it, may still hope to experience a new sensation [in Petrolia]. If he is not moved by what may on any day be observed, let him wait a little while for a flood or a conflagration!"[35] To visit such an attraction was one thing; to live each day within this place and survive under the constant insecurity was an entirely different matter.

Oil's proclivity to burn, more than the swindlers, prostitutes, and drunks, made residing in Petrolia difficult. A *New York Times* correspondent captured the landscape of oil when he said that "the ground is apt to be forested with derricks, shanties, tanks . . . all saturated with petroleum and only awaiting the falling of a spark or the scratch of a match, to blaze up with fury. Added to this, many parts of the surface are creamy with the all-pervading liquid, which settles on a thousand little pools of water or trickles lazily toward the nearest creek." The author then added the alarmingly flammable gas seeping from the wells throughout the region:

> Now, suppose that, at the moment when oil is struck, when there is apt to be a rush of gas and petroleum to the surface, a single spark from the engine or a tobacco-pipe comes in contact with this ascending column, what can the consequence be but a conflagration? In a single second the blaze may leap upward fifty feet, seizing upon derrick, engine-house, and every other inflammable object in the neighborhood. If the day be windy, and plenty of other works are standing close by, the probability is that hardly one will escape.[36]

If one resided in the valley, the proximity of crude threatened the lives of the entire family.

The lack of care discussed previously caused most of the danger; the industry could have been run safely if it had chosen safety over profit. Instead, journalists were struck by the "rudeness and imperfection" of the technology being applied in Petrolia. This writer went on to observe that "if ever there was a missionary field for intelligent engineering, it lies not far from Oil City." The call would, for the most part, go unheeded. One Oil City fire in 1866, for instance, destroyed the major businesses along Centre Street at a loss of $1 million and left 50 families homeless.[37] Fires came often to cities that took little care to prevent them. The major effort to promote safety in Petrolia came from random signage urging "NO SMOKING" (fig. 5.2).

In June 1866 the *New York Times* reprinted a description of an awful fire in Oil City. The article reported that many buildings were destroyed, but it also made careful reference to methods used to fight the fire. While the firefighting methods were not revolutionary, the writer expressed surprise that merchants and townspeople had for once chosen to fight a fire. Residents cut loose barges along the banks of Oil Creek when the fire neared so that they would not convey the fire to the other side. "A large fire pump placed in the creek . . . together with a steam pump attached to the tanks, kept the flames at bay and saved the entire city; also an immense number of lives from destruction."[38]

Fig. 5.2. United States Petroleum Company Office, Pithole

During 1866, the mention of Petrolia's litany of fires had become so ubiquitous in the popular press that papers took an inventory of the year's fires in
July.[39] Losses were placed in excess of $1 million total in the sixteen "major"
fires over the previous twelve months. The article concluded with neither an
explanation of why fires were so common nor a list of new modes of fire pre-
vention and control. Petrolia existed as a place where such fires were always
possible, and this was the threat under which those in this valley had to live
during the 1860s. Yet there were other dangers and difficulties, too.

During the 1860s, deforestation, road and derrick construction, and poor
water management accentuated the flood and mud problems inherent in the
Oil Creek valley (figs. 5.3 and 5.4). Located on the flood plain of Oil Creek,
Oil City is literally bisected by the creek. During the boom, it endured more
flooding than any other oil town. The most distressing floods, however, had
to be those for which nature could not be blamed. For instance, March 1865
brought great rains and raised Oil Creek to levels that saw four feet of water

Fig. 5.3. Road problems, Oil Creek, 1863

on many Oil City streets. As the rain continued, however, barges of oil loosed themselves upstream and flowed down the river until they reached the bridge at Oil City. The barges pressed against the bridge; then debris of all sorts added itself to the impromptu dam. Oil Creek literally rerouted itself and poured through Main Street in western Oil City, producing flooding unlike any seen previously. It was estimated that a hundred buildings were afloat in the creek.[40] The waters swept away the majority of the structures, with estimated losses at more than $2 million in real estate and oil.[41]

In Oil City, the local newspaper recounted that such tribulations in everyday existence "seemed to add determination to the heroic men, striving as they were, to make Oil City the business center of the oil regions." The writer estimated that Oil City's losses by fires, floods, and bank failures between 1861 and 1866 were over $4 million.[42] And yet the population continued to skyrocket. People came, survived, and, for a time, constructed a life and culture from what they found.

Fig. 5.4. Mule or horse mired in deep mud/crude mixture

When the oil rush swept through Oil Creek and further down the Allegheny River to Franklin, townspeople leaped to recall which of their neighbors' drinking water tasted *most* like oil. While the petroleum industry defined itself and its overall processes around the boomtown, a few towns managed the boom while keeping their community structures intact. Franklin and Titusville, Pennsylvania, offer very different stories, but each exemplifies the extractive communities that can be created when the common resource is at least slightly controlled and regulated. The control over development rarely took the form of legislation or ordinances; instead, cultural channels were often used to exert interior economic restraint or informal community control mechanisms. One 1860 observer compared these two established towns when he wrote of Franklin: "On account of the excitement, this old dilapidated town suddenly acquired new life. As might be expected a full tide of visitors is rolling in; claims are taken with avidity. Many are at a loss to decide which will eventually take the lead in the oil trade—Franklin or Titusville."[43]

Although Titusville existed prior to the oil discovery, its population stood at only 250. It did, however, possess some community roots in churches, a school, and a fairly active lumber and iron industry.[44] Titusville contained only a few businesses when Drake arrived, including the boardinghouse in which he stayed. Although the town developed permanent status through the oil industry, during the 1860s it frequently resembled a boomtown.

Twenty to twenty-five miles from Titusville, down the Allegheny from the Oil City juncture with Oil Creek, lay Franklin. The county seat of Venango County, Franklin approached the oil boom with a long history and many previously existing reasons for being. Located at the juncture of French Creek and the Allegheny River, Franklin had been home to Paleo-Indian encampments, native tribes, two colonial forts, and, if all this weren't enough, it also had that crucial local cultural heirloom—a George Washington story.[45]

As with other frontier outposts, the availability of services and goods brought settlers to the area. Franklin developed the iron ore and lumber resources in its surrounding hinterlands and eventually became a major trading point on the Allegheny River. Some of these settlers took time to consider the strange oil that pooled both on land and in water throughout the area. John Seetin, a surveyor, provided the first written record of such practices on October 20, 1799: "I have had the pleasure of exploring the oil fishery. The manner in which they gather the oil is very curious. They first build a dam of stone and that serves to retard the passage of the oil in its floating on the surface of the water; then warm the same over the fire or in the sunbeams until it becomes about lukewarm. The water by this means settles to the bottom of the vessel and so the oil is extracted." [46]

This stunning choice of terminology ("oil fishery") immediately links the product of the seeps to the valuable oil being taken from the whale fishery. Seetin then discussed the known uses for the substance and postulated that "if the physicians of our country [England] only knew of the oil they would send across the ocean for it I am sure."

As Seetin's writing demonstrates, residents of Franklin identified the odd resource fouling their water and soil but made it simply a part of regional lore—part of the sense of their place. They concentrated their efforts on constructing a stable, multifaceted community. This infrastructure allowed them to mold the opportunities presented by the oil industry into long-term benefits for the town.

Titusville lacked such strength at the outset but certainly did not follow the boomtown cycle. Throughout the rapid shifts of the boom years in this valley, these two towns were anomalies because they differed from the standard development and collapse model. At times, each community seemed to lust for the immediate gratification of boom development, but in the long run each established a commitment to managing and controlling the industry that would take place within its boundaries. The different tacks chosen by each community offer verification that more than one path could lead towns through boom to permanence.

Franklin's *Venango Spectator* served as the region's only newspaper in 1859. While in many boomtowns one could learn of little other than recent developments in the industry, the *Spectator* chose to continue to focus Franklin's concerns on national politics. The newspaper's remarkable coverage during this early period kept readers abreast of political developments that would lead to the Civil War, but it altogether resisted mention of the development of a nearby industry.

Early on, word of mouth alone spread the oil news through Venango County. For some, the news brought instant enticement. A few inmates are reported to have broken out of the Franklin jail and fled to opportunity in Petrolia. Mechanics, farmers, and curious folks of all sorts began streaming to the area from throughout Pennsylvania, New York, and Ohio. But Thomas Cochran, the newspaper's editor, continued to find more important occurrences to rail against in the *Spectator*. He and others appeared to look upriver with a wary gaze. The *Venango Spectator* refused to mention Drake's oil well until September 21, over twenty days after the discovery. The greatest event in the region's history, and possibly the greatest industrial event of the modern era, and the local, four-page newspaper elected not to cover the story. Even more astounding, when it did finally mention the event, it simply picked up the first article that had been written about the event, that from the *New York Tribune* of September 13, 1859.

Cochran's action seems without logic. His willful ignorance could suggest, as some writers have postulated, that locals were too concerned about everyday community happenings to worry about the activities of a lunatic. They appear to have been disbelievers in this path toward progress. It is also possible that Franklin was jealous of developments upriver and that Cochran desired to wait until Franklin had oil news of its own. The main reason, however, is more interesting and revealing than either of these possibilities. The belated coverage teams with later events to make it very obvious that Franklin neither doubted the veracity of Drake's strike nor missed its significance; instead, its leaders had a distinct desire to steer the development of this opportunity toward what they identified as the long-term benefit of the community.

Cochran carefully constructed any mention of oil in subsequent issues. He resisted the journalistic model that invites open speculation and the economic boon that it brings; instead, he followed up with the October 4 announcement of the formation of the Franklin Mining and Oil Company. Finally, the October 26 *Spectator* (fully one month following Drake's strike) carried an original piece of journalism concerning petroleum. The long article gave a history of the substance's use in the immediate region and made an obvious effort to ensure that every local resident possessed full awareness of the region's special wealth prior to Drake's strike. Appearing to make an attempt at thwarting haphazard development, Cochran's article stressed that "it is not probable that a shaft sunk at [just] any point in this region would be successful." Like a classroom teacher, Cochran now seemed to attempt to instruct his readers on acceptable conduct in the pursuit of oil.

Yet the tone of Cochran's account fluctuated, as if under contradictory impulses. Just as Cochran suggested development, he changed his tone. Cochran's article stressed that Drake's well appeared limited only by the power of the pump bringing the oil up to the surface. For the reader who continued to have any doubt about what he or she should do as soon as moving his or her gaze from the newspaper, Cochran broke it down simply: "oil can be produced at a mere nominal cost—certainly less than ten cents per barrel. And as the oil is now worth one dollar per gallon in the Eastern markets, this would leave a liberal margin of profit." However, in the article's conclusion, Cochran took back the reins of progress to offer an organized way that some might wish to search for oil. In Franklin, he reported that a company of forty investors was commencing operations and that he "[could not] doubt their success." In conclusion, Cochran sent out a challenge to Franklin: "No better opening can exist for men of capital and enterprise, than is to be found here at the present time. Who will embark in the enterprise? Who will strive for the golden harvest?" There is little doubt that Cochran and Franklin had opened the town to oil development; however, by pausing in his introduction of the topic and by so carefully constructing the October 26 article, Cochran apparently attempted to place certain limitations on the town's development.

Fig. 5.5. Franklin, 1860s

On November 16, Cochran reported a strike in Cherrytree Township that produced eight hundred gallons a day. "This may appear incredible," he reported, "yet is nevertheless true." He listed other strikes along Oil Creek but altogether ignored Franklin's own E. Evans, who had independently begun to sink his own well just outside of town and along the Allegheny River (figs. 5.5 and 5.6). In this issue of the newspaper, other businesses for the first time mentioned the oil boom in advertisements. A Pleasantville department store stressed that the "oil fever" had not kept it from "bringing on an increased stock of goods."

Beginning on November 30, Cochran would write of a great number of undertakings and successes in the Franklin area, including local land sales to oil speculators and the costs of tools and labor for anyone to drill a well. Finally, Cochran would relate that a New York company was likely to found a refinery in Franklin. It is, however, the December 7 issue that contains the most intriguing coverage of the lot. Over *three months* after the discovery of oil and following many articles related to its development in the surrounding region, Cochran printed an original story announcing Drake's discovery.

> We have been possessed for some time of information respecting the discovery of [oil], but have not entered into particulars about it as the principal parties in the interest requested that we should not until further developments—or until it should be a fixed fact that Oil was produced in sufficient quantities to warrant the working of the Spring; but . . . the "Seneca Oil Company" have

made the discovery, and after working the Spring for nearly two months it has now become a "fixed fact" that Oil of a most excellent quality for illumination and lubrication at the rate of from 1000 to 1200 gallons per day is being gathered from these lands in Pennsylvania.

It is difficult to locate Cochran's exact agenda in issuing this announcement. To do so, one has to recall the activity other than pumping the Drake well that occupied members of the Seneca Oil Company during these eventful months: every effort was being made by those involved in the company to buy any land and lease opportunities with potential for oil development.

It seems most likely that Cochran was under the influence or in the pocket of the Seneca Oil Company and had resisted any specific announcement in order to stave off a land rush long enough to allow the company to purchase the lands it desired. This article's conclusion seems to support this argument: "We are happy to know that the [Seneca Oil] Company is located here in our city, and that it is our citizens who are the 'lucky ones,' and that citizens of

Fig. 5.6. Franklin, 1870s

Fig. 5.7. Grant House, Franklin

New Haven have the honor of making and developing the discovery." It appears, therefore, that Cochran was under strong contradictory pressures as editor of the region's only newspaper during this historic episode. His response was to remain loyal to local investors and to promote a model of economic development that would keep Franklin's oil fever from escalating into a terminal disease (fig. 5.7).

Petrolia had no other towns like Franklin. The boom in oil overwhelmed most communities, redefining the locales' existing culture and landscape. While Franklin did not emerge unscathed, the control with which its elite classes managed the oil opportunities ensured that newcomers could find industrial working opportunities and not the chance to purchase lands that would compromise the community's stability. Titusville, however, had no such desires of control at the outset. Oil boom meant people, of which it had few, and people likely meant development, of which it had little.

Titusville and the rest of Crawford County soon found that the oil boom brought a great deal of additional opportunity, even without significant supplies of crude. This boom often involved shifts in traditional industries. In the case of Crawford County's "other" important industry, lumber, census data reveal its continued importance as the county made itself one of the lumber suppliers for the new industry in the neighboring county. In Crawford, the actual number of establishments dropped from 124 in 1860 to 104 in 1870, but shifts in technology and need made the scale of the work entirely different from undertakings prior to 1860. Hands employed increased from 307 to 597, and the annual value of the products jumped from $273,431 to $1.6 million. Some industries lost; others such as lumber gained. Services such as the making of boots and shoes and flour and meal production diminished in the number of establishments, even though the products' value rose.[47]

While a large foreign-born population indicates Titusville as a type of boomtown, it is important to note that except for the decade of the 1860s, Titusville's growth was not explosive; instead, this growth was consistent with a permanent town that lacked a stable economy, even if it was largely dependent on the oil industry. Franklin, on the other hand, was larger at the outset but never attained the size of Titusville. In 1850, Franklin's population stood at 936 and it increased modestly to 1,303 in 1860. Finally, during the region's largest growth period, Franklin's population only increased to 3,908. Of this 1870 total, 595, or 18 percent of the total, are reported to have been foreign born and 170 African American. This places Franklin below the percentage of foreign population seen statewide in 1870 but slightly above the 13 percent for Venango County.

The change was monumental in each of these communities; yet each one, albeit in different fashion, persisted. In the end, the lesson of each one's endurance is clearly its dependence on social hierarchies and devices to formally and informally manipulate the community's rise to sustainability. For each town, the oil boom presented a great challenge to any dreams of permanence, but also the possible means to its desired end.

Often, the population shifts were only indicative of more basic shifts in everyday life. During the oil boom, Franklin's population grew by 200 percent—well above the state rate of 21 percent. And, as the percentage of foreign-born persons demonstrates, the composition of the population also significantly changed. But the biggest change exerted by the boom in

Franklin was the way that the majority of the townspeople earned their livings. There would still be professionals of different sorts, but now there was also need for mechanical labor.

Professionals or laborers, the oil industry influenced both in Franklin. Col. J. P. Hoover was representative of many living in Franklin. In late August 1859, Hoover, the Venango County prothonotary, or court reporter, found himself at the American Hotel in Titusville. He heard the tales of Drake's work, and when he returned to Franklin he arranged financing with a few others to drill further a water well along the Allegheny that he knew was rank with the taste and smell of oil. In late November he struck oil at 220 feet.[48] His wealth changed, but not necessarily his life.

Titusville did not have such local professionals to become its oilmen. Oil speculators were strangers who came in droves to this small town, which served as the major service hub for the industry from 1862 to 1868. These locally owned services came in the form of hotels, saloons, and stores, as well as eight refineries, iron and piping works, and even the factory for the nitroglycerin torpedoes used to start wells later in the decade. Ultimately, though, local capital could start individual enterprises but lacked the community foundation to start the joint endeavors that had been instrumental in Franklin's development. For instance, while schools and churches were soon established, a newspaper—which had been such an important control mechanism in Franklin—was not established in Titusville until 1865.

Once it took form, the *Titusville Morning Herald* helped in maintaining law and order in the boom community. This came in reaction to the town's self-proclaimed worst summer of "undesirable characters," such as pickpockets, gamblers, counterfeiters, and horse thieves. Titusvillians felt that they had no control over their community whatsoever. In early August, the newspaper called Pithole, the largest boomtown in the region, a "disgrace" in morality and cleanliness, but "not far behind Titusville in sanitation."[49] While it may seem demeaning for Titusville to compare itself to a boomtown, it should be understood that no one in 1865 believed Pithole was a boomtown that would disappear. In fact, some might have thought the comparison to be delusions of grandeur.

A national cholera scare in 1865 finally pressed residents of Titusville into action. A series of editorials appeared in the new *Herald* to force local residents to take control of their community. The editorial about the "national pestilence—cholera" illuminates the dangerous details of their own communities:

> The festering filth and rottenness accumulated in mid-summer is in itself sufficient to originate a devastating plague, and so far from any measures being adopted toward remedying this condition, we observe that these loathsome deposits are being constantly augmented.
>
> Look at the green and stagnant pools poisoning the atmosphere in every quarter; the sweeping of garbage and offal from eating stalls and saloons; the slimy reeking deposits that line the bank of the creek—and above all the slaughter house, standing within 10 yards of the business center, and sending its rank and fetid exhalations into every parlor, kitchen and dormitory—and then ponder on the effects of such a visitation predicted![50]

This exhortation for action continued through the month of August in accounts connecting the cholera epidemic in London with the basic shortcoming of boomtowns: a lack of interest in taking action toward permanence. "Are we not culpable then," reads one editorial, "individually as well as collectively—if we neglect to make the most of our advantages, and suffer the present physical and moral pollution of our town to remain unchecked and unrestrained?"[51]

A boomtown took no time for this sort of self-examination. Titusville's residents now began to exert control over its development through the employment of community committees. Soon, details of boom life as commonplace as the muddy streets were criticized—even the main thoroughfare of Titusville was thick with mud (fig. 5.8). This required the construction of elevated boardwalks, but even these did not prevent all contact with the mud. An August 28 *Herald* editorial called the mud at the main street crossing in town a disgrace, claiming that it was four inches thick. Combining the human element with such details, an August 29 editorial described the "strangers within our gates" as the criminal element that had arrived with the oil business.

The physical structures of Titusville reflect the ebb and flow of boom priorities with an interest in permanence. Figure 5.9 shows the residential portion of town at the beginning of its boom in 1864. The scattered fine homes and church structures in the background are overwhelmed by the bulk of the town's residences in the foreground. While these are made of wood and contain many carbuncle additions, the layout possesses order. In the right foreground, the shuttered house appears to have a young orchard in its sur-

Fig. 5.8. Mud, Main Street, Titusville

rounding yard. The fences prevalent throughout the scene limited straying of owners' livestock but also figuratively and literally protected homes and families from the boom life.

Unlike Franklin's buildings, Titusville's many wooden structures were not built to last, though a few appeared quite ornate, such as the Crittenden House and the Bliss Opera House. Unlike other boomtowns, Titusville additionally possessed a few brick structures that showed the community's desire for permanence. The Corinthian Hall, Academy of Music served as one such structure. Constructed of brick, the hall had tall windows with second-story styling around each. The roofline possessed similar decoration, and the tall plate glass windows of the first floor offered the building's contents to passers-by. In 1865, Titusville even added a small local college that was affiliated with the Episcopal Church.[52]

By 1873, Titusville, with its boom behind it, had blossomed further. Figure 5.10 shows the orderly but highly industrial scene of the town. These ac-

Fig. 5.9. General view of Titusville, 1864

tivities were located in one section so that the rest could develop as a community. This demographic control then allowed for the construction of the grandest homes in the region in another section of the town. One of these belonged to F. S. Tarbell, a local barrel and tank maker. He purchased the materials that had formerly been the Bonta House Hotel in Pithole and from them constructed the fine home in which his daughter, Ida, would spend her teenage years.

For Titusville, development and eventual stability derived from the trickle-down of profits from oil. The town became the hub of many rail lines, pipelines, and toll roads. It would develop its own Oil Exchange and become a headquarters for trading and speculating. These undertakings kept individuals (most of whom were away from home and in need of even the basics of everyday life) passing through the town, and they could then support the local service industry. Unlike the boomtowns, Titusville served purposes that did not expire with one field's supply of oil.

Fig. 5.10. Titusville, 1873

Community came to mean something different in Petrolia during the 1860s. This chapter has suggested new community priorities and also revealed how dangers and horrific events failed to move oilmen to change the way their industry functioned. While communities ultimately had little control over the development of the oil industry, they were able to muster at least a nominal united defense against larger structural changes. Those living in Petrolia most often united against threats to human labor, which moved the region to join together to find solutions to shared problems. This tendency is evidenced by the numerous organized uprisings of teamsters and other workers following each technological development in transportation and storage throughout the 1860s.

These strikes and uprisings followed the construction of plank toll roads and pipelines, but none had the intensity of the large-scale reaction to the 1872 discovery of the monopoly known as the Southern Improvement Company. Arriving at the moment when other regions had displaced Petrolia as the nation's largest producer, this fight additionally stands as at least the symbolic last gasp of individualized speculation in the valley. The personal chance that defined the economic culture of the 1860s boom faded as corporate structuring loomed on the horizon. The progression from Blood Farm to the Southern Improvement Company illustrates the industry's development during this period. The success of the company fed the demise of the other faction of oil's society, but the culture of the new industry had created each.

Led by the company Rockefeller and Paine out of Cleveland, the Southern Improvement Company neither owned property in the oil regions of Pennsylvania nor leased derricks. Each of the members of the company lived elsewhere in the United States. From their various vantage points, however, these businessmen administered all of the railroads over which the crude could leave Petrolia for points north, south, east, or west and the major refineries at which the product would be processed. In essence, they held the Pennsylvania supply in their own control like a tap that could be opened and closed to the rest of the world. Oil producers in the valley banded together and elected not to sell their commodity to Southern. This, of course, meant that they sold to no one.

As the data discussed above relate, no resident of Petrolia remained unaffected by the oil industry. This was a battle that concerned every resident. However, it was also a manifestation of their own creation—of the laissez-faire industry they had allowed to prosper and take hold. Like Dr. Frankenstein, they needed now to watch the monster run its course. The *Times* correspondent wrote that "it will be found that a class of men accustomed to risk every dollar they possess in sinking oil wells will show themselves capable of making any sacrifice to repel" this conspiracy to take from them over $30 million annually. Soon, national coverage referred to the episode as "the Oil War." Refiners in other areas also supported the laborers and small producers of Petrolia against Southern, who planned to make Cleveland and Pittsburgh essentially the only refining points for Pennsylvania crude.

Without contracts for transportation, Oil Creek refiners shut down their operations and halted most oil production. After forty days, they had destroyed the monopoly and production began again. Within months, however, production had produced such a glut of oil that the oil producers voted to shut down their wells for a month during the fall. Those involved in the industry had learned the positive outcome of banding together. One could call such a moment the high point of the society of Pennsylvania's oil boom, because such instances of cooperation remained rare; however, the fundamental reason for cooperation was profit from oil. Crude had come to dominate every facet of life in the Oil Creek valley, as well as to serve as the basis for nearly every community or individual priority.

The triumph over Southern proved short-lived. Rockefeller, the wolf, lay in wait. The spirit of the teamsters had actually played directly into his "Plan," which he would implement later in 1872. Hiding within the Southern Improvement Company for much of the late 1860s, Rockefeller laid the groundwork for his effort to control the entire industry. Rockefeller formed the Standard Oil Company of Ohio in 1870; by 1879 Standard controlled 90 percent of the U.S. refining capacity, most of the rail lines between urban centers in the Northeast, and many of the leasing companies at the various sites of oil speculation.[53] Throughout the 1870s and early 1880s, Standard Oil would further its dominance over the refining industry and by 1882 had formed the Standard Oil Trust.[54] Within such a corporate framework, Rockefeller made his fortune while squeezing out the spirit of the early boom. He made the industry fit his vision of it: oil towns became "work camps," and the only towns in the process, boomtowns.

After a decade of incredible economic boom, the incorporated industry now shifted with the latest hot spot. Of the 1860s and 1870s, historian Daniel Yergin wrote, "The substance for the popular form of lighting [kerosene] worldwide was provided not merely by one country, but, for the most part, by one state, Pennsylvania. Never again would any single region have such a grasp on supply of the raw material."[55] The state's dominance would continue for a while longer, although the Oil Creek valley would lose its distinction (see table 5.2). First, wells from the lower Allegheny River area usurped the top producer position in 1872. Then followed Clarion County, other portions of the Allegheny River district, and Bradford County.

The culture of the 1860s boom along Oil Creek would be re-created in miniature wherever oil towns sprang up, but the completeness of oil's domination over life in Petrolia could never be duplicated. The title of this chapter derives from an 1866 description of Oil City. The writer speculated that if development continued, this region and specifically the town of Oil City would become all that nature had intended. Every development in this valley during the 1860s began with the preconception that the human being is capable of discerning nature's intention better than nature can itself. It ended in a cultural landscape defined by industry and corporate life and a cultural legacy that repeatedly continues to create such locales. But the places pass, and it is more accurately the commodity that becomes all that humans, not nature, dreamed it could be—and much more.

No name was given to this drawing board city. Some people called it Holmdenville, some Pit Hole City, others just Pithole. It had no legal status, no incorporation. Pithole was a private development isolated in near wilderness.

On May 24, 1865, after noisy fanfare, five hundred town lots were offered at public sale. A thousand buyers were on hand to get in on the ground floor, crowds huddling around huge maps showing lot numbers for . . . prospective renters.
—William Culp Darrah, *Pithole: The Vanished City*

Chapter Six

Pithole: Boomtowns and the "Drawing Board City"

The man steps from the stage and into the muddy morass of Pennsylvania's latest industrial undertaking. He strides from the stagecoach's prop and onto the planks that surround the business offices of the United States Petroleum Company—his company. He has come from New York City to see it all for himself. Once securely on the wood platform, safe from all but the flying globules of mud, he turns and takes in the scene. A steady stream of noisy wagons pulled by balding mules and horses squeaks and squeals past with its burden of barreled crude. The animals struggle through the knee-deep mud but are always forced to take that next step in the right direction—forward.

Mr. Evans watches the teams pass, each one sinking until the wagon's axles skim the top of the thick mud covering the site's main thoroughfare. He is invigorated with pleasure and confidence as he views his company's latest development along Pithole Creek. During the rest of this day, November 9, 1865, Evans will gleefully view the nearly completed plank road that enables teamsters to take his product more easily to Titusville for a trip down Oil Creek and beyond. Most amazingly, however, he will also see the six-inch pipe that will eventually take oil from the world-famous Frazier Well to the rest of the world without the use of teamsters or horses. From every angle, he will view progress. For many years to come, the oil industry will base its technology on the successful efforts of his companies and others like it.

The sights of this day made Evans so optimistic that he remained unfazed later in the day when he witnessed the Frazier cease its flow.[1] Such developments paled when compared with the twenty-four tanks full of crude and $45,000 worth of buildings and equipment confronting Evans at the U.S. Well. The locale clearly was less a cultural place than an industrial process. Developers soon realized that the significant flow of oil, its remote location, and nationwide interest could create another profitable money-making process as well: town speculation.

This extractive site called Pithole soon gained a national reputation as the greatest of all boomtowns and the symbol of the progress embodied by the developing oil industry. From an industrial work camp, this place suddenly warranted comparisons with the great cities of the Northeast—and it was even claimed that it would become Pennsylvania's second city after Philadelphia. This was an astounding transition from the farm that had formerly filled this space. Within six months of the discovery of oil here, a town had taken form—the largest boomtown in the region—with a population of ten thousand. This population would spring from zero to its reported high of fifteen thousand in less than eight months.

Following his November visit, Evans issued the United States Petroleum Company's first annual report. His report leaves no doubt about his own convictions about this commodity's future and the role that Pithole was to play in this development. The report claims that the construction of Pithole in the last six months points to the conclusion "that petroleum, a great gift of God, for man's benefit, held in store for ages, and recently given to us in our day of national trial, will not vanish but continue through time to give a good and cheap light in the houses of the poor, lighten the burden of the taxpayer, increase the national wealth, be useful in the arts and manufactures; add a page to the volume of scientific discovery, and flow a steady stream of profit into the pockets of these interested in its production."[2] To its developers, the place was only part of a process in achieving its sought-after product. As such, it formed a town type that would make Petrolia famous.

As we have begun to explore, towns in Petrolia developed in a variety of ways. Constructed around oil, very few permanent towns grew out of the new industrial process. As we have seen, a small number of previously existing communities were able to fit the industry into their existing social structure; however, there is a much longer list of other towns that during the 1860s existed for a matter of years, months, or even just days (see Appendix). These were boomtowns: attempts at communities that grow out of economic boom.

Whereas Oil City was the quintessential outcome of Petrolia's industrial process, the early oil industry also created a systematic process to act as a cookie cutter for almost any other locale possessing oil. The boomtown rested at the core of this system of extraction. While the industry did not go on to produce dozens of Oil Cities, oil exploration globally continues to be riddled with the remnants of boomtowns. The boomtown became an economically efficient mode through which large companies could transmute even the remotest portions of the globe into their system of extraction. Propped up for a moment, the boomtown could support the industry and its workers while they engaged in oil extraction. The entire developmental process of such towns became entwined with the industrial system of the oil business until they were one. The boomtown became a necessary part of oil development—an outpost that could be placed in Midland, Texas, Casper, Wyoming, or Talara, Peru. But it all started in the Oil Creek valley.

Best understood as a developmental device within the oil industry, boomtowns were designed with no interest in community or long-term life. This phase within the larger industrial oil process was, in fact, perfected as a standard type of development in the Pennsylvania oil regions after it had been used in dispersed examples in the American West. The boomtown became a process by which companies could make great sums of money, not just on the resource, but also on the chance to look for that resource. Similar to purchasing all the stools in a casino of slot machines, the company held control of one's

Fig. 6.1. Shamburg, 1868

ability to play "the game"—in this case, the game of speculating in oil. One had to subscribe to the company's plan to even attempt to reap benefits.

Initially, the oil boom towns seemed to occur almost organically. This process was presented for everyone in the *Titusville Morning Herald*'s coverage of the development of Shamburg in 1867. The correspondent introduced the area of elevated land along the Oil Creek: "comprising several farms and tracts, each is a distinctly defined oil field by itself, yet throughout the entire region, the territory in this vicinity has been and still is known as Shamburg, a title that at once belies its seeming falseness when we consider the value and amount of petroleum."[3] Instead of illustrating a sham of any sort, Shamburg represented a "striking example of what the steady perseverance of one energetic operator may accomplish by the development of what was at one time considered valueless territory" (fig. 6.1).

Shamburg, where the first well was struck by Dr. G. S. Shamburg in 1866, was on the site of the former Sheridan Farm but did not retain the agricultural name. Price depressions in the market held off the growth of the field and town until a year later, in 1867. When strikes warranted additional wells and leases in an area, enterprising individuals would attempt to convert ex-

Fig. 6.2. Home of Dr. G. S. Shamburg, 1860s

isting structures into service facilities or would develop an entire multifaceted center for commerce—a town. The *Spectator* article appealed to widespread public interest by including a section on the "moral status" of the town and those living in it. From 1869 to 1870, the article reported that there were forty-eight assault and battery charges levied and fifteen illegal sale of liquor charges. Otherwise, though there was a great variety of crimes (twenty) for a small town, the number of infractions rarely topped five. Shamburg, one and a half miles by stage to the nearest railroad connection, possessed one church, two schools, one post office, and three telegraph offices.

Recording the holdings of the town was an attempt to enlist new investors and inhabitants for Shamburg. This was part of the culture of Petrolia: occupants were always talking about the next frontier of oil. New Shamburgs sprang up overnight, only to be abandoned during the 1870s when the supply of oil was depleted. Yet such transience still constructs a culture. In Shamburg, for instance, figure 6.2 shows the simple wood-sided I-house that was home to the area's most prominent citizen, Dr. Shamburg. While the home structure is solid but not suggestive of opulence, the well-dressed women present a rare view of high-class ladies in Petrolia. The industrial structures can be seen directly behind the home. Most interesting is the tiny derrick next to the child on the side of the house. This appears to be a toy miniature of the derricks one sees throughout the Oil Creek valley. The toy provides a glimpse of the culture that sought to disseminate the industry across family generations.

It was around 1870 that a science for planning such boom communities was defined. Whereas most of these towns had been previously constructed by individuals in fits and starts and had followed some sort of traditional New England townscape whenever development remained constant over a few years, central authority of ownership and administration began to influence the planning of these towns near the close of the decade. In towns such as Red Hot (fig. 6.3), it appears that the planners had resigned themselves to

Fig. 6.3. Red Hot, 1870s

the transience of their community. Red Hot appears more like a settlement along the frontier of the American West than the previous boomtowns such as Shamburg. This was an extractive landscape without delusions of permanence or of community. Red Hot was a work camp, which used nearby Shamburg for its commercial needs. Yet there is no denying that in visions such as Red Hot we see a landscape whose use has been codified, stratified, and, most importantly, planned.

Made clear by each of these examples, a culture of boom management developed among business owners, some transient, others based in Titusville, who rented commercial space in the new boomtowns. More than any other detail, population shifts determined the boomtowns of Venango County. Cherrytree, site of one of the runs (small streams) most popular in the early years, saw its population increase between 1850 and 1870 from 930 to 2,326, and decrease by 1880 to 1,618. Oil Creek Township went from no population in 1860 to 5,098 in 1870, before falling to 526 in 1880. Cornplanter boomed from 1,000 occupants in 1860 to nearly 10,000 in 1870. And Pithole City, the prototype extractive community, rose from nonexistence in 1860 to over 10,000 in 1865, fell to 237 in 1870, and was not even listed in *Census 1880*.

The development and history of Pithole tries our traditional definitions of terms such as town, community, and even landscape. Indeed, the place compares most easily to factoryscapes where each person and business relies entirely on the creation of a single product. In essence, the town's reason for being is unequivocally based on a single commodity. Is this then truly a town? Does it have any of the traditional earmarks of a community? And if it is devised on a drawing board by an economic developer interested only in extracting oil from the earth, can it even be called a culturally created landscape? Yet it is through exactly this testing of definitions that the boomtown phenomenon allows a penetrating foray into the ethics and values of the industry that took shape in the Oil Creek valley during the 1860s.

Cultural geographer John Brinckerhoff Jackson has written that "no group sets out to create a landscape. . . . What it sets out to do is to create a community, and the landscape as its visible manifestation is simply the by-product of people working and living, sometimes coming together, sometimes staying apart, but always recognizing their interdependence." He continued, "It follows that no landscape can be exclusively devoted to the fostering of only one identity."[4] Under Jackson's logic it would seem that a community is incapable of existing where it is organized under a single motivation. No built landscape better exemplifies this logic than the boomtown, particularly those that remain so completely dependent on the single entity under which they have been organized that they cease to exist when it is exhausted. The oil boomtown as a type fits this criterion very well, and possibly none so clearly illustrates the extremes of the process as does Pithole.

Once the developmental process had begun, the oil industry knew only one speed: fast. Pithole's petition for incorporation, submitted in September 1865, tells the details of a site where four months earlier no signs of civilization were found.

> In view of the wonderful and marvelous growth of Pithole within the past two months it has assumed the proportions of one of the most respectable and flourishing towns in this Commonwealth.
>
> Although laid out not more than two months ago Pithole now contains over 300 houses, ten large hotels, fifteen or twenty blacksmiths and mechanics shops, two banks, a post office, a variety of other business establishments and over 2,000 inhabitants.
>
> Over ten million dollars worth of property is invested here and hundreds of thousands of dollars are being invested daily.[5]

Holmden Street very quickly became a main commercial thoroughfare. Merchants and investors were so certain of the town's permanence that they aggressively competed for the most choice building sites. Building lots continuously and repeatedly changed hands as prices skyrocketed, often multiplying the original price by four or five times after only two weeks. This ethos created an industry and a culture in all of the boomtowns that would spring up; however, Pithole distinguished itself from all others.

The nation quickly became fascinated with Pithole: in that way the town is a fitting representative of development, the rule of capture, and the potential of myth. As William Culp Darrah wrote: "Pithole was unique—not in the sense that there were no other oil boomtowns but in the sense that it was a microcosm of American inventive genius. Pithole literally played a key role in the rationalization of the infant oil industry but it was much more than that. Pithole attracted to its environs the ambitious, the adventurous, the greedy and the wicked."[6] Pithole presented a model of boom that became acceptable, if not desirable. Often, writers described the place grandly, similar to the presentation of a small town developing in the American West.

Once visitors arrived and rid themselves of the imagined descriptions, they often learned that they had been misled by the promise of a stable, civilized town.

A few early visitors, however, sought not to deceive. In actuality, Pithole possessed more than enough to intrigue readers without embellishment. J. J. Bouton wrote for the *New York Journal of Commerce:*

> There is a transitory look about everything in Pithole, the houses are knocked together as if not meant to last but a season and the streets are allowed to remain unpaved and unplanned. . . . Nobody thinks of trying to improve the streets. Is there not a symptom of misgiving and mistrust in all these temporary arrangements? Does it not really look as if the people thought their oil might give out, leaving them in the forlorn condition of Rouseville and Plumer which were growing like gourds before the Pithole strike and which were abandoned in consequence of it?[7]

The appearance here derived directly from each town's dependence on a finite resource for its entire existence. Seen throughout Petrolia, the boom-town became the apex of the rationalization process on the landscape of the Oil Creek valley. The roots of this ethos were based in the ability of the industry and its components (human and mechanical) to move on with the supply to new boomtowns wherever the next strike occurred. This ethic of transience became the prototype for the American boomtown and would soon dominate extractive industry in general.

Pithole's name took on new meaning when the town boomed to a pit of immorality and grime; yet the actual meaning predated the boom and grew out of the regional legend of three hunters passing through the area. When they stopped to rest by an outcropping of sandstone only a short distance from a creek, the hunters noted a foul smell, which then led them to notice deep fissures in the rock. One climbed down the cavern a bit and then returned to sit on the edge with his comrades. Upon doing so, he suddenly fainted. After reviving him, the three pronounced the holes exits from hell and called them Pit Hole.[8] The name, later merged to Pithole, has also been attributed to the early troughs or cribbed pits that settlers found in the area and attributed to Paleo-Indian groups. In any case, there is no doubt that the region was known as Pit Hole prior to the area's oil boom.

Production in the Oil Creek valley had been so great during the early 1860s that markets could not absorb it. This glut caused a sharp drop in the price, which then led to massive abandonment of towns and leases in the region. While oil now was overabundant, experiments in illumination and lubrication promised new markets soon. These new uses merged with a widening of international markets to revive the industry. Finally, as 1865 began, profits exploded and massive amounts of foreign capital rushed to the area.[9] Petrolia stood poised on the edge of economic boom. To get over this cusp, though, it needed a magnificent attraction to arouse national interest.

With the infusion of investment, speculators began exploring areas away from Oil Creek. During this 1864–65 push they pressed nearer and nearer to the two Holmden farms along Pithole Creek. Leases were taken out from the Holmden brothers and other surrounding farmers in late 1864.[10] The lessees then banded together to form the United States Petroleum Company. Two of these individuals, I. N. Frazier and James Faulkner Jr., had been employed at the Humboldt Refinery, located in the nearby town of Plumer. Frazier and Faulkner were on hand in June 1864 when Thomas H. Brown, who was divining with a witch hazel twig, selected the precise spot for their first well. Before drilling the United States Well Frazier died of a heart attack. The other investors chose to rename the soon-to-be-famous well the Frazier. His death caused a halt in the work, and drilling ceased until July before continuing through the fall.

By Christmas, the Frazier Well had been drilled to six hundred feet, which was much deeper than the wells along Oil Creek. William Culp Darrah observed that "the organizers of the United States Petroleum Company cared little about finding oil. They were selling stock. Oil made no difference. What they had to do was drill a well to satisfy stockholders."[11] While this is an accurate assessment of the company's motivation, it also must be appreciated that the discovery of oil would allow the company to accrue much more profit from stock sales.

On January 7, 1865, Faulkner arrived to find the well producing 250 barrels a day. The call went out from the telegraph office in Plumer and echoed down the valley: "Deep oil had been struck along Pithole Creek!"[12] The remote location of the well immediately forced Faulkner to struggle to construct storage tanks and arrange transportation. The March 2, 1865, *Oil City Register* presented a vision of the scene:

What a strange and busy scene around the well. A number of men are hewing trees, hauling logs and making corduroy roads to render easier the ascent to the tank. Other men are putting finishing touches to a huge storage tank to hold twelve hundred barrels of oil. . . .

Mounting a rough ladder, you get your first view of the oil which has been so rudely disturbed from its long slumber far down in the very bowels of the earth. You see nothing but a two-inch iron pipe, with a stream of fluid flowing out as large as a heavy hydrant stream, and looking like countryhouse molasses, and of about the same consistency.

Every couple of minutes the gas which can be plainly seen issuing from the tube-like waves of heat, gives the stream a little spurt, and then it resumes its even and steady flow. Standing over the well, the oil can be plainly heard ascending the tubing. Nearby stands the engine house, with its trim, polished and powerful engine from New York, looking as demure and innocent as if its ceaseless and powerful workings were not the cause of all the hub-bub. No need of an engine now, except to sink another well, for this "big well" is a flowing not a pumping one. It runs by nature's gas, not man's steam. How long it will thus flow, who knows?

In this description one finds the machines of the age. Amazingly, nature—in the form of natural gas—actually assists the industrial process so that the process seems natural. In truth, nature played only a small part in Pithole's oil boom (fig. 6.4). This boomtown, like others, quickly became a completely industrialized, extractive community.

Pithole's first well had been struck and, truly, timing was everything. Similar strikes had been made in the Oil Creek valley during the first five years of the oil boom, and boomtowns took shape around them in order to provide the goods and services that would be needed. However, during the early months of 1865, thousands of soldiers were discharged from the Union Army. These men flocked to the most likely source of jobs. As if staged as an act in a play, Pithole burst onto the scene and represented the greatest possibilities available in the entire nation. Pithole was suddenly poised to boom as no town ever had.

Oil companies needed to attract a workforce to the area, but the improved economic situation and increased outlets for crude had also made companies more willing to risk large sums of money on oil. The first boom for Pithole was an explosion of oil companies: by February, ten oil companies had registered in the area; capital certification for companies in New York, Massachusetts, and Pennsylvania had been registered at $350, $160, and $145 mil-

Fig. 6.4. Pithole, 1865

lion, respectively.[13] Many of these companies were complete frauds that never drilled a well or actually sold a lease; however, they advertised in newspapers throughout the nation and were more than happy to accept investment. Darrah summed up the situation concisely: "Land speculation, not oil, was the means to quick profit."[14]

It took approximately $15,000 to secure a lease, erect a derrick, drill six hundred feet, and meet expenses for the four or five months required to find out whether a well would succeed or not. Most developers chose to spread this risk among a number of investors. If a private investor had enough capital to do it all himself, he would still most likely invest it in as many as fifteen wells. By mid-August, just prior to Evans's visit, a correspondent for the *New York Tribune* climbed up the rigging of the Frazier and counted more than three hundred derricks on the Holmden Farm and adjacent land.[15] Darrah speculated that one out of every eight wells in the Pithole field yielded oil, but only one in twelve in an amount that paid.[16]

By the end of the summer, three thousand teamsters were in the area to drive the oil out of the Pithole area by wagon.[17] The only buildings in the immediate area were the homes of the two Holmden brothers and a log cabin known as Widow Lyon's house. Such a setting left the hordes of visitors and workers with no place to eat or stay. The Holmdens began serving meals to nearly two hundred visitors per day, with workers always eating first. In May 1865, Col. A. P. Duncan and George C. Prather, two businessmen from Oil City, purchased land adjacent to the oil fields. They were not interested in speculating for oil; their interest lay in creating a town to profit from the great sums being reaped by those speculators residing in the area temporarily.

After purchasing the land, Duncan and Prather laid out the "drawing board city" and began letting leases. Even the first leases for property in the town reflect the predominance of the ethic of transience. Leases for a store or business cost $50 for six months or $100 per year, renewable for three years. This price allowed the leases to move briskly, but with one catch: at the expiration of the three-year agreement, the lessee had the privilege of removing the buildings that he had constructed or selling them to the owners of the land at the landowner's figure. Even if a business owner thought he or she might not be in this place permanently, they may have wished to have a sturdy or nicely kept building; however, with the knowledge that anything they built would basically return to the landowner from whom they rented, few worthwhile improvements would be made. The entire process of town settlement encouraged the tenant to think in terms of a short stay and quick profits.

The first building, the balloon-style Astor House, took one day to construct.[18] Within four days, two streets of businesses were up and running. After six weeks, two thousand white males, eleven women and one "colored person" resided in the unnamed town.[19] In addition, a thousand teamsters and laborers scattered throughout the valley depended on Pithole for services. These services usually consisted of providing alcohol and meals, which Pithole obliged with in abundance.

Like all boomtowns, the community developed in a backwards fashion. Whereas many towns settled in the West put the infrastructure in place first and then developed out of and around it, these oil boomtowns procrastinated in setting up the infrastructure in case the town did not last. For instance, no sanitation existed in Pithole. "The whole place smells like a camp of soldiers with diarrhea," observed the correspondent for the *Titusville Morning Herald*.[20] Privies were insufficient and poorly maintained. Residents simply tossed garbage out of back doors and left it to decompose. Many observers complained of the rank smell from carcasses of mules and horses discarded in the brush along the edge of roads.

Prices for food and other goods generally ran 15 to 20 percent higher than in Titusville.[21] For instance, fresh fruit, except during the apple harvest nearby, remained nonexistent at any price. By far, though, water proved the bane of Pithole's existence. With none of the surface water suitable for human consumption, it became the most valuable commodity in the boomtown. Many of the oil companies privately piped water into their buildings from the nearby hills; however, inhabitants of Pithole were forced to purchase drinking water from peddlers in the streets. The crisis grew during September 1865 when water sold for fifty cents a barrel and five cents a pail with some peddlers charging ten cents a pitcher. If only one could drink some of the abundant—and cheaper—crude! Hotels and restaurants hauled water in by wagon.

Water provided the first opportunity for Pithole to try to band together for its common good. Franchised on September 1, the Pithole City Water Company began looking for water on the hills above Morey Farm.[22] Near the end of the month, they struck water and excavated a 25,000-gallon reservoir while also laying a system of iron pipes through the main streets of the city. However, applications for use were only made by businesses. From the beginning, the Pithole City Water Company proved unable to make ends meet. Too few users were willing to pay, and establishing town authority appealed to no one—even though such authority would keep the mains full at night, when they were most needed for fighting fires.

By August, the population of Pithole had grown to around fifteen thousand. The seedy nature of many of Pithole's major establishments made law and order impossible to uphold. The vast majority of the population consisted of transient workers, almost all of whom were male. Women in Pithole most often worked at one of the town's saloons or brothels. Brothels had a long history in Pithole, including a small private home near the old homestead that became known as Heenan's Cottage, the first brothel, almost as soon as oil had been struck. By October, competitors had gathered in more central locations along Holmden Street.

The theaters, saloons, and hotels of Pithole each ran fairly legitimate businesses, but many additionally offered outlets for the work of prostitutes. After gaining cash or experience in Pithole, many of these prostitutes went on to follow frontier settlement west. Most of the establishments employed between six and ten women. Handbills distributed throughout the town advertised "girls from all the big cities" and "girls from every state in the union." One scholar estimates that by December 1865, four hundred prostitutes were employed in Pithole, where the entire population hovered around ten thousand.[23]

Imbibing alcohol made up the main pastime of most off-duty teamsters. During this era, when the town had two thousand permanent residents, there were twenty establishments selling whiskey, which was often used to seal deals. According to the correspondent for the *Nation*, who visited in 1865, "It is safe to assert that there is more vile liquor drunk in this town than any of its size in the world. Indeed, a bar is almost the invariable appendage to every building. Lawyers have bars appurtenant to their offices—each hotel, dwelling-house, or shop has its separate bar."[24] These hallmarks of immorality defined Pithole's national image. Individuals such as Ben Hogan and French Kate achieved renown in pulp writing throughout the nation. Each one ran establishments of uncertain morals, and Hogan evolved into one of the oddest characters in Petrolia by following his famed boxing matches and boxing school/saloon in Pithole with the kidnapping of a young woman. He eventually became a born-again Christian and took off on the lecture circuit, bringing Pithole even more publicity.

The theaters and opera halls of Pithole were a bit more respectable than these other leisure establishments. Bringing authentic national entertainment troupes into one of the most awful places on Earth was no mean feat. Such cultural pursuits differed from any the region had ever before witnessed, but suddenly a stop had been added for major performers traveling the state. As a footnote to history, John Wilkes Booth performed in these halls immediately prior to his trip south to assassinate President Abraham Lincoln. There are many unconfirmed reports of individuals overhearing his rantings about the Civil War and Lincoln, but no one reports hearing of his plans prior to April 14, 1865, when he took the life of the nation's heroic leader.

Booth and other similar characters needed only a temporary bed to call home. The 1865–66 *Pithole City Directory* reported fifty-four hotels in the town of Pithole, an additional twenty-one within a one-mile radius, and another five or six within a three-mile radius. Similar to the Astor House, most of these were built from green timber; as the wood dried, large gaps were left between the boards, making these hotels and boardinghouses infamous for their draftiness. The dried, unpainted wood also became highly flammable.

As the earliest hotel available, the Astor often squeezed two hundred guests or more into its forty rooms and served a thousand meals a day. The United States Petroleum Company soon converted its employee quarters into Pithole's second hotel, the United States. The original utilitarian structure had cost $5,000 to construct; the renovations and furnishings added to make it appear a plush hotel cost upwards of $20,000. The oil company immediately began leasing its hotel for an annual rate of $30,000. At this time, nightly rates were around $2, and the earning potential of the hotel was estimated at approximately $100,000 a year.[25] Other establishments popped up, literally, in the form of large tents. Most often, the tents filled needs temporarily while construction took place. During the summer, two grand hotels opened to try to appeal to the high-class trade. The Morey House was opened as the grandest hotel ever seen in the town. It even boasted a fresh spring of its own to bring water to its patrons. Ownership problems, however, over both the hotel and the oil leases additionally held by its owners forced the Morey House to close in the fall of 1865. It stood empty for the entire Pithole boom and was set on fire in October 1866.

The Chase House had a more lasting presence in the town. Set up similarly to the main hotel in a frontier town, the Chase House presided over

Fig. 6.5. Pithole, Duncan House in background, business office in foreground

Holmden Street with its grand furnishings and popular billiard room and bar. In addition to the saloon, the first floor contained the Pithole Post Office and the office of the Union Express Line, from which stages left twice a day. On the second floor, Western Union Telegraph maintained an office. In addition to accommodations for five hundred, the Chase offered meeting rooms, a ladies' lounge, and a small library.

The only true competitor for the Chase was the Duncan House (fig. 6.5), which was a similarly well-furnished, four-story structure. The Duncan, however, also incorporated gas lighting, steam heat, and a newly patented oil stove in each room. The saloon and dining area catered to the idea behind the Duncan's construction: family accommodations. This would be the place where families could stay while men carried out oil business, or so the planners intended. Planners soon learned why this niche had not yet been filled: no families came to Pithole. If any came to Petrolia, it would be considered foolhardy for them to travel en masse to Pithole. The Duncan barely stayed alive for a few years and never made enough to repay construction costs.

Most often, developments in the boomtown followed necessity. Life in Pithole cost a great deal of money overall, and accommodations were the primary expense. Most workers went to rooming houses instead of hotels. Soon, the fine hotels had either closed or degenerated into brothels of one type or another. Pithole grew so rapidly that suburbs even began springing up. Each one followed the model of Pithole: temporary houses and buildings went unpainted and often unfinished, and restaurants and other establishments strove only to be cheaper than the next.

Holmden Street in Pithole became the entire area's main street (fig. 6.6). The stores, like all the other buildings, were entirely incongruous with one another and most were poorly built.[26] Very few buildings possessed paint or other adornments. These were luxuries of another world, another culture. Many times, the high cost of construction and the temporary nature of the businesses pressed merchants to group together in one building. Up to four businesses could often be found under a single address. Darrah describes Holmden Street as a "hodgepodge of improvisation."[27]

The pace of business grew even more intense with the opening of the Pithole Post Office on July 25. By September the post office passed three thousand pieces of mail each way daily—a pace that for a short time placed the office third in the state, behind only Philadelphia and Pittsburgh. During this period, the *Pithole Daily Record* listed the names of those with unclaimed mail. Judging by the size of this list, the greatest number of transients reached Pithole in September; the women working at brothels also peaked during this period.[28] The post office's first day also marked the opening of the first bank—and within three weeks the bank's deposits had reached $250,000.[29]

Where did such funds come from? Leases on Pithole's building lots continuously resold at skyrocketing prices. Duncan and Prather, who had begun the town only two months prior, decided to sell in the fall. They had purchased the farm from Thomas Holmden for $25,000 in June, and in September, after a great deal of dickering, the farm was sold for $2 million. Holmden, meanwhile, was happily earning royalties on the thirty-five hundred barrels a day that were being taken from the farm and the $60,000 a year

Fig. 6.6. Holmden Street, Pithole, 1866

rent being paid by merchants in Pithole City. However, Duncan and Prather
still gave him a bonus of $75,000.[30]

Most surrounding farmers reacted similarly to Holmden and sold or
leased their land for development. One of the few exceptions was Parcus
Copeland, who refused to sell for reasons more of greed than of agrarian ide-
alism. A. G. Morey agreed to purchase Copeland's farm for $200,000 in
March; the two, however, filed the paid price as only $85,000. When the deed
was filed, tax was demanded on the amount, and Morey had to decide
whether to continue the lie or not. Meanwhile, great amounts of oil were
coming in from the first few wells that had been sunk on the farm, and the
Morey Farm Hotel had been opened for business. Copeland suddenly began

demanding the $115,000 balance between the two amounts (which Morey had already paid).[31] A lawsuit followed and the deal was nullified.

The *Titusville Morning Herald* tells the entire confusing story of how Morey and two other speculators traded the deed back and forth while trying to decide the best time to sell.[32] The oilmen entirely disregarded Copeland, the original owner, during this process. Embittered about the inside trading taking place with "his" land, Copeland decided to do all that he could to get the land back. To further confuse the issue, on April 12, the Homestead Well, one of the greatest in the area, came in just off Copeland's land but near enough to significantly increase the property's value. Copeland then chose to pull in one of the only women mentioned in the speculative boom— his wife. He claimed that Mrs. Copeland refused to sign the deed to her home and farm unless she was paid $75,000 (even though it was common knowledge that the initial sale had had her support, though not her signature).

During this entire period, rapid development of the farm continued. Actual ownership, it seems, mattered very little in the process of economic boom. In fact, at times it seemed unnecessary, if not foolhardy, to own the land upon which the boom was playing out. Actually owning the land would create a burdensome problem when the boom finally subsided. In the Morey case, public sentiment shifted against Morey and his partner. They were forced to admit that they had been offered $1 million for the farm—but only if they had a clear title. Copeland ultimately received $300,000. All told, Copeland eventually made nearly $1 million from the sale of oil leases; however, he continued to live in the log cabin that he had built when he first settled the land. Originally, he had kept 104 acres to farm, but that eventually shrank to his 5-acre reserve among the oil derricks.

Without ownership or community control, the environment of the boomtown became overwhelmed by users seeking only oil. The lack of basic necessities defined the oil boomtown and were of little concern to speculators. Getting around Pithole was nearly impossible. Most streets were paralleled on both sides with hand-dug ditches about eighteen inches deep. These proved a futile attempt to manage water runoff and to try to control the mud problems. A few merchants placed plank walks leading to the entrances of their shops. But by the spring of 1866 the loaded teams made Holmden Street impassable on a consistent basis. The mud was infamous and made life in the town impossible. During separate incidents, at least two women had become sunk up to the knees in mud and had to be extricated.[33]

In the spring, a few donations made it possible to plank or corduroy, as it was known at the time, certain portions of First and Holmden Streets. Residents also installed long parallel planks to serve as walkways.

The mud problems stemmed largely from the rapidity with which the forest and farmland of the area had been laid bare. Darrah, a paleobotanist by trade, observed, "mud and rock spread over the land killing everything. The oil seepage finished the job, ruining the land for decades to come. With the removal of the grass and other vegetation the ground became a quagmire of mud when wet and a thick accumulation of dust when dry."[34] But these alterations were truly of no concern to residents of a town who had no intention of making it their permanent home. This lack of concern extended through every aspect of life in the boomtown. Even though individual investment was active in towns such as Pithole, these communities remained most basically work camps designed to extract oil.

This lack of care made Pithole the oil region's most acute tinderbox. The danger of everyday life made any domestication of such places impossible and irresponsible; instead, the boomtown could be maintained only as an industrial locale. An English banker observed of Pithole, "I have spent many hours in great powder magazines yet I would rather pass a month in them than a single day by the wells of Pithole which are simply as dangerous as powder magazines without one of the precautions."[35] The intense danger of residing in an oil boomtown precluded most families from making such a place home. In truth, these locales were industrial sites that did not marry well with children. For instance, in June of 1865, when two tanks containing twenty-five hundred barrels of oil burst and flowed into Oil Creek, two boys could not resist seeing what would happen if a match were touched to the flow. They struck a match and flames soared up the valley. Teamsters immediately built an earthen dam to stop the fire's movement. This proved to be the only thing that stopped the boys' play from causing horrendous damage.

The previous incident began with a burst storage tank, a frequent occurrence in the Pithole area. After the first few disasters, local newspapers began calling for the use of iron tanks. Further accidents were necessary to warrant the extra cost to those carefully monitoring the bottom line, but eventually iron tanks were employed. Of course, the logic sought to prevent burstings (thereby securing the supply of oil) if fires occurred, not to prevent fires and leakage. In essence, the operating assumption in Pithole ran that fires would inevitably occur.

Signs adorned Pithole urging people not to smoke or light matches (fig. 5.2). Some wells employed guards to enforce such restrictions. The awesome danger of fire existed as standard knowledge for all residents; however, the industrial practices combined with the life of a boomtown made fires a common occurrence. The Grant Well came in on August 2, 1865, and led to one of the town's most famous conflagrations. The owners were surprised to get such a strong flow so quickly, and they lacked any receptacle in which to put the three hundred barrels a day that were being pumped out of the well. The oil simply flowed over the ground, filling all the gullies and depressions with raw crude. The owners would collect it later, or so they surmised.

Great excitement followed the striking of the Grant, and spectators flocked in. At 7:30 in the evening, with about a hundred people standing nearby, a great explosion took place. The flames soon covered an acre of land and rose a hundred feet. Twenty people suffered serious burns, but only one died from injuries. Most escaped injury by jumping into the creek. The man killed had been one of six sitting on the derrick when it exploded. Two months later, the well would burn again.[36] Fires and the impending threat of them were a constant awareness for those living in Pithole and similar towns.

Exceptions to this exploitation existed even in Pithole. Some residents tried to halt their town's eternal binge, yet they seemed always doomed to failure. Nevertheless, churches and religion played a formative role in Pithole. Rev. George R. Ormond conducted the first services in town in the Metropolitan Stables on July 1. Ormond moved on, but two weeks later the Erie Conference assigned Rev. Darius Steadman to gather a congregation in Pithole. Steadman's journals reflect that, overall, the townspeople were receptive to his presence. He soon took up residence in the corner of a carpenter's shop.[37]

Steadman differed enough from traditional ideas of ministers that he appealed to the rough-shod Pitholeans. He wore a heavy beard and smoked a pipe, both of which were frowned upon by the Methodist Church. For months Steadman held informal services wherever possible, often even in saloons that opened doors just for the services. By August 14, committees were in place to secure donations for the construction of a church. Personally, Steadman had a natural appeal to these people, but they also appeared to be interested in the religious instruction he had to offer their community.

Fig. 6.7. Panorama of Pithole, fall 1865

Thomas G. Duncan donated two choice lots to the Methodist Society, and more than $8,000 was secured in donations.[38] Services began in the new structure on January 1, 1866. From its hill to the rear of much of the town, the fine structure dominated the scene of Pithole (fig. 6.7): the imposing nature of the structure reflected that of Steadman himself. As a pillar of morality in a vastly immoral locale, Steadman was greatly respected and widely influential. The congregation grew rapidly and had plenty of funding, because Pithole's wealthiest residents constituted the congregation's core.

A Presbyterian Church also soon joined Pithole. Unlike Steadman, the first Presbyterian minister was all too similar to his parishioners. After a particularly moving sermon in one of the local hotels, one listener presented the Reverend Hughes with a one-sixteenth share in a producing well on the Hyner Farm. Hughes immediately negotiated the sale of the interest for $10,000, pocketed the money, and abandoned preaching.[39] Although these funds would have built a new church, the Presbyterians were left to try again.

Pithole presented unique challenges for ministers and priests. On December 1, 1866, the *Pithole Record* reported that mass would proceed regularly in the new Catholic Church built on land donated by the owners of the Holmden Farm. Later dedicated as St. Patrick's Church, the new church appealed to the largely Irish teamster population. Father Finucane, its first priest, however, always had a difficult time attracting a congregation. In his most ingenious scheme to attract churchgoers, Finucane announced the raffling of a gold watch. The raffle went so poorly that Finucane was forced to place a notice in the *Pithole Record* to explain that the watch would be raffled that evening at his home. "Let all who have got tickets to sell, try to be honest enough to return me either the tickets or the money and seek not to swindle the Church as some have done already."[40] By May, Finucane had moved to Franklin and the Pithole parish had closed.

Other attempts at community control did not include religion. The *Pithole Daily Record* began publication on September 25, 1865. The press legitimized the business community through advertising and the town itself by organizing happenings into one cohesive unit. The paper would rise and fall with the rest of Pithole, but it did have one distinguishing accomplishment. The newspaper's masthead presented the town's name as "Pithole" instead of the "Pit Hole" used previously.[41] As its first community-wide reflection of itself, the newspaper presented the news that would generate a feeling of community in the boomtown. To this end, its spelling of the name would be automatically grasped by everyone involved.

These symbols of the formation of a standard community, however, were always confronted with the reality that Pithole existed as an oil camp, exceedingly dependent on the oil laborers. No group made this reality more difficult to escape than the teamsters who transported oil to market along the awful roads of this remote spot. Driving wagons pulled by mules or horses, these men were predominantly Irish. Driven hard, the animals were often unsightly due to oil's proclivity to remove hair if the owner did not clean it off.[42] The drivers generally lived in boardinghouses, ate all their meals in restaurants, and enjoyed spending time in saloons. These service businesses were the most dependent on the teamsters' support; however, the teamsters' work was also one of the most critical cogs in the oil process, which supported all of Pithole. From this area, six miles from the nearest town and without access to navigable waterways, the poor roads served as the only method of transporting the oil.

The Pithole teamsters practiced the ethics seen throughout the Oil Creek valley to an extreme. They continually raised the fee for transporting crude by wagon. Whether the oil was destined for Titusville or the Allegheny River, the fee rose quickly from $3 to $3.15 per barrel. This gave the teamsters more profit per barrel than the operator got for producing and selling the oil. And when roads were most difficult, the teamsters refused to travel them. Teamster greed more than anything in Pithole fueled technological experimentation. In July 1865, private companies created a plank road to Titusville and instituted toll charges for traffic on it. Eight million board feet of heavy oak and hemlock timber from the surrounding hills went into the road. Coopered together by William Webb, a Nantucket shipbuilder, the planks were four inches thick and laid on a solid foundation.

These roads involved a great deal of maintenance and did not at all increase the number of trips a teamster could make in a day. Bad weather caused many delays on these two-lane roads; in addition, traffic and distance still restricted teamsters to a single trip a day. The teamster rates as well as the slow speed of transport created a great need for storage tanks in Pithole. Most of the oil was pumped directly into casks made of wood (fig. 6.8). The profusion of tanks filled with standing oil, of course, significantly enhanced the danger of fire in Pithole; however, the corporate priority remained to pump all the oil out of the ground before someone else could get it. Once in a tank on its property, the oil was safely under the company's ownership. This interest again belies the boomtown's delusions of permanence; these towns were more oil process than place or community.

Due to the continued difficulties with transport, 1865 saw oil interests continuing to struggle for alternatives to teamsters. Two projected railroads began a race to completion in late July. At about the same time, calls began for the construction of the world's first oil pipeline to carry crude down to the Allegheny River. Some oil businessmen voiced concern that once all three transportation alternatives were in place, none would be productive. Yet with each under private development, the new modes were seen as moneymakers in their own right. Each simply pressed to be the first completed.

By October a line of 2-inch pipe had been laid in a trough stretching from Pithole to the Miller Farm on the Oil Creek Railroad line. A total of three pumps forced through 81 barrels of oil an hour, which equates to the one-day total of 300 teams working a 10-hour day. Developers added a fourth pump, and if run 24 hours a day, the line was capable of moving 2,500 barrels a day.

Fig. 6.8. United States Well, showing open storage tanks

They then laid a second pipeline along the same path. Generally, oil was transported at $1 per barrel, which compared favorably with the teamster charge. The pipelines caused an uproar among the teamsters, forcing five hundred out of work in the first five weeks. Teamsters threatened the developers, sabotaged the line, and even set fire to some of the storage tanks of one company. Finally, with no other alternatives, teamsters cut their prices to $1.25 a barrel.

Pipelines and the other technologies also contained a basic weakness. The grand pipeline of the Titusville Pipe Company best demonstrates this problem. Planners engineered a new line from Pithole to Titusville to serve as a permanent portion of the industrial process. The costs for its construction far exceeded any of the others: just the ditching, carried out with great care, cost $1,000 per mile; force pumps throughout the line cost $800 each; and the steam boiler cost $2,000. When completed in March 1866, the pipeline plan's one "given" had failed them: Pithole's oil supply dwindled. By the time of the line's opening, the entire daily production of Pithole made up only half of the great line's capacity.

Fig. 6.9. Pipeline dump site, 1866

A roof overhung the area in which users would dump their oil into the end of the pipe (fig. 6.9). The wagons were backed onto a one-foot-high embankment so that the barrels could then be rolled down a chute and emptied by hand. Such a system still required some teamsters, but they needed only to make repeated short trips. This led to the 1866 development of an accommodation pipeline that bypassed teamsters completely. The accommodation lines ran directly to each well and linked each to a central storage area. From this storage area, oil could immediately be pumped miles through pipelines and loaded directly onto boats. With Pithole as its nexus, the new process for producing and dispersing crude had evolved and become standardized.

Technological developments such as these were normally not carried out by trained engineers and were rarely patented. The oil industry desperately lacked professionally trained scientists and engineers. Methods were adapted from other industries and improvisations completed when necessary.[43] A lack of such trained professionals—who would have functioned outside of the speculative process—is sorely obvious in the early industry, and specifically in the construction of places such as Pithole. The final development, later in 1866, in the transportation of oil was the railroad tank car. These cars offered the cheapest method of conveyance yet.

By Christmas more than a thousand teamsters had left Pithole. The pipelines left them with few alternatives. This exodus crushed Pithole's commerce and hopes for permanence. Businesses began closing and services were streamlined after only five or six months in existence. Such a fate befell any boomtown. No effort was made to diversify industry in hopes of supporting the suddenly unemployed: industrial rationalization would never allow such humanitarianism to penetrate the utilitarian logic of Petrolia. Pithole existed as the major supplier of oil, and the industrial process took shape to get it out of this remote place—if one of these cogs failed, the entire operation would collapse. Like a victim attached to life support, the livelihood of every occupant of this town derived directly from black gold.

Pithole reached its peak production in October 1865, reportedly between six thousand and eight thousand barrels. On the Thomas Holmden Farm there were eighty-one wells on seventy-nine leases, eleven of which were producing and three only marginally. These same statistics show that the entire Pennsylvania oil region estimated production at nine thousand barrels—of which Pithole alone produced at least six thousand. Of the Pithole supply, over half came from only two wells. In a place where the product had become the rationale for every development, these two wells sustained the largest town in the oil regions.

Even during its October peak, few voiced concern about wells running dry. Yet that would only be one of the problems facing Pithole's effort to endure. In December 1865 through January 1866, Pithole experienced one fire per week. Figure 6.10 offers the amazing sight of just one of these, the Duncan house (seen earlier in fig. 6.5), in flames. Finally, one thousand people gathered in Murphy's Theater and called for the installation of a few hydrants in the town and a fire brigade.[44] Arsonists, with a variety of motives, started most fires. The town tried to quell this practice by instituting a lynch law. But Pithole had no ability to cope with a large fire or even to notify its occupants in the event of one's occurrence.

In the end, lack of interest and the inability to rally any sort of community sentiment would thwart attempts to stabilize the town. By spring 1866, the fire company had disbanded, giving up due to lack of support. Throughout the rest of 1866, Pithole experienced one fire after another. In one June blaze, twenty buildings along the main streets were lost. Another fire in early August swept through the oil fields. In all, estimates of loss over this period of time were placed as high as $3 million.[45] It seemed impossible that the town could continue—and indeed it could not.

Fig. 6.10. Pithole on fire, 1865. Compare to figure 6.5

People had left Pithole en masse starting in late 1865. Business owners attempted to sell the grand hotels and found no takers. They held raffles to dispose of the fine properties, and again found no takers. Finally, most were abandoned and others sold for scrap. By January 1866, the population had fallen to barely four thousand. Then the oil supply began giving out as well. In February 1867 another fire destroyed almost all of the remaining businesses in Pithole. The oil regions began a "Pithole Relief Fund," and Horace Greeley came to speak on the town's behalf. The *Daily Record* discontinued publication in July 1868. The flash that was Pithole was over. Founded on a drawing board in May 1865, the town had boomed and completely busted in little more than three years.

Ironically, the years that Pithole enjoyed success were too short to appear in the census records. In *Census 1860*, Pithole, of course, did not exist; in *Census 1870*, the town's population was down to 281. Since Pithole had fallen between the cracks of the decennial census, the story of its boom was left to be told by those who had lived it and observed it. In August 1877, the Court of Quarter Sessions revoked the borough's charter. Pithole was struck from the face of the earth, and ceased to exist.

What would Mr. Evans think if he had stepped from his carriage in 1895 and viewed the scene that John Mather captured in figure 6.11? Most likely he would glance about and consider it a "clean" job. Fire had erased almost any record of the industry that had thrived there and the community that had simply tried to exist alongside it. He would view the site with the clinical perspective of a practitioner of industrial logic: the oil had been extracted effectively, and the industry as well as its labor force had moved on to the next site—the next boom, the next Pithole.

But what of the human story here? Had this not been a town? A community? What of the fate of those played out here? The lives of those involved? All these things, in the mind of Mr. Evans and others, did not reflect the bottom line. Such ideas or concepts were drained of meaning and significance by the process of industrial boom. The boomtown heaped human culture and society, along with the meaning of specific places, upon the growing pile of waste that had now become an inherent part of technological progress.

Despite the efforts of Evans and his compatriots, meanings of place are not static entities. The meaning of a place can be redefined or reclaimed by later observers. The sites of Civil War battlefields are one example of this. A place can become a landmark when something that it represents becomes of

GENERAL VIEW of PITHOLE, in August, 1895

Fig. 6.11. View of Pithole, 1895

critical cultural importance to later generations. The sites of battlefields represent sacrifice, valor, and the power of ideas and beliefs. Their sacredness
derives from one's ability to stand in the locale and reflect upon the action
that took place there; however, it also grows out of the power of hindsight
and one's ability to consider additionally all the related issues and ideas that
have transpired throughout the nation.

Pithole, too, has taken on meaning for today's observers. It is now the Pithole National Historic Site. With the sites of former streets cleared of tall
grass, and with photos to show what it once was, this former boomtown—
lacking community meaning, by the definition of J. B. Jackson—takes on renewed visibility and, with this, renewed significance. The site has been transformed from meaninglessness to great importance. The site of Pithole, in its
current state, becomes a place again, but with even deeper meaning.

Jackson and others discuss the idea of "sacred landscapes" as those places
that have been imbued with a deeper cultural significance than what is immediately apparent from mere physical appearances. Pithole's docile present

contrasts with its uproarious past to create a somber reminder of the costs of industrial development in terms of human life and culture. It has become a sacred landscape that can remind each viewer that humans live in a delicate balance. It is a balance that is upset not only by natural disasters such as fires, earthquakes, or floods but also by the more subtle parts of our lives such as our cultural values, priorities, and ethics.

When Jackson wrote the following words, he imagined a hypothetical site, possibly a hybrid of many. His insights, however, act as an epitaph for a landscape that never fit his definition of community: "[The] condition of being part of nature brings with it certain responsibilities and restraints. To damage a system which allows an infinite number of life forms to coexist, to destroy what we cannot possibly replace, would not only be irresponsible, it would threaten our own survival."[46]

The visitor is often able to perceive merits and defects in an environment that are no longer visible to the resident. Consider an example from the past. Smoke and grime badly polluted the industrial towns of northern England. This the visitor could easily see; but local residents tended to shunt unpleasant reality out of mind, turning a blind eye to what they could not effectively control.

—Yi-fu Tuan, *Topophilia*

Nature is a very careful accountant. . . . every change in one part of an ecosystem sooner or later has some effect, however minute, on every other part. Human beings and their industries are no less part of the ecosystems in which they work than are the plants and animals they harvest. Human economies, however, account for the costs and benefits of their activities through the market mechanism, which unfortunately is much less efficient as a transmitting medium than an ecosystem is.

—Arthur McEvoy, *The Fisherman's Problem*

Chapter Seven

Delusions of Permanence

On October 28, 1865, the *New York Times* chose to put the Civil War behind the nation and move on with other matters—specifically, future commerce. In its major cover story, the *Times* called upon the nation "to concentrate all [its] energies for the development of [its] material resources." Many resources and sites could occur to the reader, but to the writer one predominated. "Of course, men's eyes rest naturally for a time upon that wonderful region in Western Pennsylvania which has contributed so largely to our revenue during the past four years."[1]

The article explained that this realization led others to organize a trip to introduce Petrolia to all those monied people who had been preoccupied with the war. These invitations were carefully targeted to potential investors (which the paper listed at great length), but also to another constituency. The "grand excursion to the oil regions," as it was called, also invited writers from newspapers in nearly every major U.S. city for a week-long junket in Petrolia. The *Times* greatly increased the pool of those aware of the trip by including in its article a full copy of the invitation text—as if inviting the nation to join in.

> For the purpose of affording the capitalists of the country an opportunity of inspecting personally the Pennsylvania Oil Regions, an excursion through them has been arranged, in which you are invited to participate.
>
> Two hundred guests will be in attendance, including prominent business men from each of the principal cities in the Northern States. Everything will be arranged to make the trip as comfortable and pleasant as possible, and to afford a full and thorough view of the wonderful phenomena of the oil regions.
>
> Your expenses during the entire trip and return will be paid by the projectors of the excursion.

Suddenly, status had become part of Petrolia's mythic attraction. The list of invitees seems intended to salute those few but also to anger the many left off the list. In fact, Petrolia's coming-out party had been organized by one man with a very specific purpose in mind. The "inspection" would be focused on one site—the gleam in the organizer's eye, the newest oil town, Reno.

Petrolians' yearning for permanence in a temporal industry had found new energy after the Civil War. They began to think that they could control their own fate and divorce themselves from dependence on the commodity that had brought them there. These aspirations to permanence can only be viewed as ironic. While the meaning of a locale always derives from a variety of social, cultural, economic, and geographical characteristics, Petrolia's meaning overwhelmingly derived from its product and national image, both of which were generated externally.[2]

As we have seen, the details that made residents' lives difficult became hallmarks of the American view of industry. These details, most of which grew out of the industry's transience, also quickly came to dominate this locale's sense of place during the 1860s. Not only was this sense being created by external observers, but the place—as well as its towns and industry—was never meant to endure. One of the region's greatest users, John D. Rockefeller, captured the sense of the place when he disdainfully referred to the

Table 7.1. Outside Involvement in the Oil Regions, 1865

Place	No. Companies	Capital ($)
Philadelphia	286	164,475,000
New York	121	128,950,000
Pittsburgh	63	21,610,000
Boston	14	6,800,000
Baltimore	8	2,250,000
Brooklyn	3	1,250,000
Cincinnati	4	1,150,000
Erie, Pa.	2	400,000
Chicago	2	350,000
Cleveland	3	930,000
Other places	18	9,400,000
Now forming	21	19,000,000
TOTAL	545	356,565,000

Source: *Venango Spectator*, Feb. 22, 1865. From Pennsylvania Historical and Museum Commission, Drake Well Museum Collection, Titusville, Pennsylvania.

chaos and disorder as a "mining camp." The landscape of early oil would not be made into the efficient, corporately designed locale that Rockefeller's order-obsessed mind craved. It was meant only to be exploited and then cast aside, while the orderly industry moved on.

Most problematic, however, these externally imposed points of view defined this place's meaning and significance in a lasting way (table 7.1). While this was one thing when the boom raged, it is quite another when this image continues to define the place that remains today. In other words, not only were the resources of this place harvested by outsiders, but the discourse to define the region's sense of place was also dominated by national culture—specifically, the ambition for economic and technological progress. The endurance of this appropriation is staggering when one considers that today this place's meaning still derives from the oil boom of the 1860s.

These realities and their enduring lessons require me to focus once more on this sacrificed region and the way that the public perceived it during the

1860s. A place so exploited had no future, and this precarious position attracted the public to tales of life in the oil industry. Yet the region needed to learn this lesson once more, since Reno allowed the aspirations to permanence to become full-blown delusions.

Prior to the Reno trip in 1865, the mythic aura of the oil landscape had clearly captured American interest. Representations reproduced oil scenes for stereoptical viewers and later in the form of postcards (figs. 7.1 and 7.2). Petrolia even became a destination for travelers and thrillseekers. One writer observed that an 1865 trip to Petrolia made him "satisfied that it is one of the most interesting and wonderful phenomena on the face of the country, altogether independent of its financial and social aspects."[3] When the scene of extraction became a spectacle to people of the 1860s, one journalist wrote, "One can hardly be said to be well posted as to the resources and occurrences of this country, unless he has visited in person the petroleum yielding tract of Pennsylvania."[4]

In the 1860s, Petrolia became similar to a tourist site, composed as much of what people wanted or expected to see as of its own distinctions. In this fashion, one can view the culture that redefined this place as one of tourists or visitors, observers detached from the place. Such individuals possess specific perceptual tendencies, described by cultural geographer Yi-fu Tuan: "only the visitor (and particularly the tourist) has a viewpoint; his perception is often a matter of using his eyes to compose pictures. The native, by contrast, has a complex attitude derived from his immersion in the totality of his environment."[5] By this standard, few of the land users in the Oil Creek valley were natives: their perspective most resembled that of the visitor, even if they might stay in this locale for months or years.

Since it lacks personal reference, it is common sense that the visitor's perception of a scene is more shallow. Philosopher Paul Shepard writes that the tourist "moves in a sphere which has no immediate connection to the conduct of his daily business. He observes the pattern of stream life in which lives the fish, and the whole watershed may assume some significance. . . . He moves through novel landscapes which have a minimum of stereotyped sign value."[6] Shepard refers to this as the state of the "itinerant eye." The itinerant eye allows one to be unbound by the limitations of local or regional meaning and culture; instead, in a postmodern perception, the viewer is freed of all but his personal or culturewide preconceptions.

Fig. 7.1. Stereoptic view card

Fig. 7.2. Colorized postcard scene

The culmination of such observation is often a construction of cultural tastes and attitudes, or what Shepard refers to as "scenery." In the attraction of scenery—in the decision that something is appealing—we see a culture applying aesthetic value. Mental definitions of revulsion or attraction often derive from art. In this fashion, knowledge of places raises people's interest, but pictures make the sights into scenery.[7] Observers normally link the term *scenery* to the visually appealing; however, it also applies to views that are distinct for any number of reasons. The landscape of Petrolia during the 1860s was certainly distinct, and its oddity stirred something in Americans reading of or viewing it. The meanings that viewers derive from any type of scenery can potentially tie deeper emotions to a specific view. Philosophers of the eighteenth and nineteenth centuries postulated that an overwhelming natural object could attract human attention by first making the viewer aware of his own limitations and then awakening a sense of human worth when the viewer realized the ability to conceive of the impressive object. As unbelievable as it may seem, during the 1860s the scene of Petrolia stirred similar emotions in the American public.

The sublime, thereby, becomes an emotive response to an aesthetic. In this fashion, a type of depiction can become indelibly linked with an additional, subliminal feeling and emotion.[8] For this to be possible in a view of the physical landscape, the natural world first had to be seen in more than a utilitarian light. The romantic movement was one of the earliest shifts in these sensibilities, freeing many in the Western world to perceive the natural world with wonder and not fear. As many of us have learned from the writings of Leo Marx and others, the nineteenth-century observer did a strange thing with the sublime. In possibly the greatest example of their overwhelming trust in economic progress and technological development, American viewers imbued scenes of industrial progress with meanings of sublimity.[9]

These feelings of awe generally contradicted the aesthetic scenery of such places in written or illustrated descriptions, which were often fairly unpicturesque. Now, for instance, included among the sublime were icons such as a gristmill, which identified a mode of progress and community success through technology.[10] Examples such as Petrolia reveal that the technological sublime did not strictly stem from the dichotomy of the pastoral and the industrial, as Marx describes. Both socially and aesthetically, technology could represent a source of excitement and not a disruption of any natural order.[11] In fact, the technological sublime often helped onlookers to see massive technologies as merely an extension of the natural order—the progression of humans toward economic development.

The technological sublime was a neat marriage but, in aesthetic terms, a coexistence that could not possibly survive the rigors of time. Natural beauty would inevitably become blighted at the hand of human technology. Yet people were attracted by the energy and potential that such a place exhibited, and this allowed the technological sublime to go through levels of refinement during the nineteenth century. Early on, an attractive scene required that technical accomplishments be set among picturesque natural scenes. In this fashion, the borderland between the human and natural worlds could be made to look beautiful and progressive at once. Soon, however, the nineteenth-century American became so enraptured with development that natural splendor need not accompany material progress. In fact, natural blight could even be substituted for the former beauty and the scene would still be emotive.[12]

No matter how squalid and unpicturesque, a sight remained capable of evoking the technological sublime and thereby serving as a model of the positive capabilities of human technology. Even if nature had not been entirely tamed, the process of human technology exerting itself upon natural forces attracted the itinerant eye. This human-created nature often seemed wonderfully wild and out of control.

While the human dominion over nature initially attracted Americans to such scenes, views such as Petrolia did not necessarily overwhelm the onlooker by exhibiting nature under human control. Often, nature remained in control of these scenes; or at least the human element did not effectively control nature. Indeed, at any moment nature could strike the entire scene from view. The attraction of such a place, therefore, becomes the process of gaining control through the contest between two forces, the human and the natural. The site becomes an active border between these two worlds.

In the scene of Petrolia, the natural landscape of the Oil Creek valley resisted human domination with fire, flood, mud, and many other weapons. For visitors, neither the natural nor the technological alone stirred the imagination. The combination of these factors—the ebb and flow of human technology's ongoing struggle to control this resource—attracted observers. As such, human failures and their ghastly results took on popular appeal.

While Petrolia would never have significant tourist appeal, details of the extractive landscape certainly aroused the American itinerant eye. As the Civil War relinquished its hold on American popular life and writers began to travel to Petrolia and write of their observations, the new horizons of the industry gained definition. Americans were enthralled with the seed that had germinated in this place while they were at war. The money that had changed hands for lease rights and property ownership in the early 1860s appeared staggering, yet it proved nothing compared with the sums that would be exchanged if national sources were tapped for long-distance speculation.

The desire for extensive long-distance involvement led to one of the most revealing episodes in Petrolia's history. The attempt to establish the town of Reno makes up the industry's most significant attempt to found a community in the same fashion as it did a boomtown—by design. The trip came during a brief window of time when even Petrolia could host tourists seeking "scenery." The inevitable end to the trip and the boom reconstruct the story as a parable about the meaning of place in human life.

By 1865, Petrolia had become an entirely industrialized locale, in which every component—human or natural—was measured and weighed only for its relation to the swift generation of capital through oil. By the time of the

close of the American Civil War, this region presented itself as a type of fac-
toryscape, where, with wildly varying degrees of efficiency, industrialists and
speculators had devised a technology to get oil to the world. Yet the industry's transient nature continued to distinguish it from others. With aspirations to appear more like iron, steel, and even coal, a few developers added a
new agenda to petroleum development.

In the mid-1860s, a few industrialists decided that the oil business needed
to be more stable and permanent to achieve greatness. Until this point, of
course, its impermanent appearance had been one of the oil boom's claims
to fame. This attempt at repackaging would, in fact, represent an interest in
harking back to earlier ideas of shared industrial and community development: views of communities in which the town controlled the industry for
its own purposes. The complete failure of this effort demonstrates Petrolia's
distinction from earlier industrial development and even places it in a position of defining much of subsequent American industrial development.

Whereas the early 1860s had seen inhabitants of Petrolia reveling in the
mythic oddities that distinguished their locale from any other (thereby
arousing popular interest and possibly investment), the mid-1860s introduced a need to substantiate the industry's secure standing. In this fashion
the industry sought to shed the ethic of transience that permeated its every
detail and to reconfigure itself as a long-term, stable venture. The method
for doing this, however, remained the Barnumesque methods that had established its mythic aura of danger and gross excess in the first place.

Leading this charge was Charles Vernon Culver, who, after one successful well strike, had begun buying the region's banks until he owned a total of
thirteen. He soon became renowned for his financial control over the region
and industry and also for his ability to raise money when necessary. His ad-
vice alone, it seemed, could lead to the success or failure of a boomtown. In
1864, Culver gave in to the pressure of his supporters and became the
youngest congressman in the U.S. House of Representatives. It was an occupation that he had not particularly desired, and he rarely attended sessions.
Throughout his tenure in politics, Culver's heart and ambition remained in
the oil regions.

He soon focused his efforts on making Franklin, the town in which he
resided and held the most sway, the next great oil town. This act of ego was
obvious to all, because Franklin lacked the single most basic component of
an oil boomtown: great amounts of oil. Franklin produced only a scant

amount of oil and had few flowing wells, compared to the Oil Creek valley. In fact, as was discussed in Chapter 5, the town had little interest in being such a place. Investors in the processing town ridiculed his attempt and refused to support it. Culver nursed a grudge over the episode. His reaction, in typically Petrolian terms, was to flex his influence elsewhere.

Soon, Culver had purchased twelve hundred acres along the Allegheny River just below Oil City and designated it Reno, the next center of the oil industry. The developer set his sights upon ending the industrial centrality of Oil City, which was the base for many of the individuals who had rejected his efforts to develop Franklin. Oil City had the advantage of being located directly at the mouth of Oil Creek, and any oil descending the creek by skiff had to pass through there. For this reason, Culver set out to use a developing technology to his and Reno's advantage.

Culver immediately hired engineers to construct a railroad from Reno to just one site. With the oil from this single locale, the town would seemingly be able to support itself and squelch Oil City's dominance. The town, of course, was Pithole, the renowned king of oil boomtowns. Culver aimed for nothing less than all of Pithole's oil to be shipped to Reno for processing and distribution. The possibilities were exciting, and Culver had very little problem raising support for the rail line.

Even though it would be an extremely expensive undertaking to develop Reno, Pithole's immense supply of crude gave Culver the confidence to use his and others' fortunes. Similar to a badge of honor, Culver flaunted the amount needed to develop Reno to anyone who would listen. It was the most preposterous sum ever bandied through the oil regions: $20 million, $10 million of which would be held to back up the town's initial development. Even Petrolia lacked this kind of money. It would require investments from people not yet involved in the region. As the Civil War came to its end, Culver seized his opportunity.

He organized a great trip through the oil regions to reveal the wonders of the industry to new investors and to spotlight his project in Reno. He would cater to the itinerant eye of America.[13] The event swiftly took on national prominence when journalists presented Petrolia as a potential destination for tourists. If we analyze the newspaper coverage of this excursion, we can grasp exactly what this place meant to its nineteenth-century observers. The first question to be dealt with is simple: Why would travelers be attracted to this awful place? First, Culver planned the trip carefully and lavished gala balls and dinners upon the participants. Most importantly, however, Culver brought the entire nation on the trip as well. The inclusion of journalists as guests ensured that his story would be widely disseminated.

The organization of this expense-paid trip is not entirely without precedent. However, the extended coverage by the *New York Times* is extraordinary; and similar coverage appeared in every major newspaper in the United States, as well as some lesser ones. In essence, thanks to the popular media, the organizers of the trip appealed to all uninvited businessmen who wished to catch the investment opportunity.

The articles produced by the writers traveling on the excursion testify to Americans' pleasure in reading of wealthy Americans luxuriously going about their lives. Many of the journalists matter-of-factly explained that the group, as one writer had it, "was composed chiefly of 'capitalists'—a favored portion of the community who are generally supposed to be so overladen with money that they are constantly running around the country seeking investments."[14] The other portion of the group, the journalists, were, in another writer's words, along "to record the history of the tour and tell the world of the realities of the fabulous tales which 'occasional correspondents' have been charged with manufacturing in their own fertile brains. . . . [The excursion] assures the daily historians whose province it is to separate the chaff from the wheat and give their readers only the genuine grain."[15]

The stops along the excursion included the most permanent, refined towns of the region, but also Pithole and some of the most memorable wells. Finally, Culver showed his hand and made the final stop the prospective site of Reno. In essence, the first days of the trip were intended to impress visitors with the financial potential of the industry, but also to revolt them with the most base details of the industry by taking them on a horseback ride along Oil Creek. Reno would then be presented as the one place offering both the promise of wealth and refined, family living. Because of the previously held mythic perceptions of Petrolia, such coverage enhanced interest nationally. Many of the articles were set up like travelogues in order to allow others to follow the trek themselves.[16] Evidently, the squalor and difficult travel in such a locale did not preclude it from serving as a tourist destination.

With varying degrees of emphasis, the journalists followed Culver's prescription perfectly. They bemoaned the roads, mud, and money being thrown about, and then concluded their installments with a description of Reno, the only oil town with delusions of permanence. One journalist wrote,

"Schools, churches and pleasant dwellings will be erected, and Reno will be adapted for the permanent residence of those interested in oil. At present the oil towns are most uninviting places. The houses are small and poor, and all social comfort is neglected. Reno will possess the features of a home."[17]

Each of the writers would make careful note of the industry's waste; specifically, the thousands of exhausted wells that stood as "monuments of folly."[18] Whether it was through Culver's specific direction, or only influenced by his intent, the writers by and large came away noting the transience of the rest of the towns in Petrolia. Of Pithole, which was by far the highlight of the trip, one writer noted that "a whale lying dry on the beach, would not be more helpless than Pithole City minus her oil supply. Banks, billiard saloons, theatres, opera houses, concert halls, and all that is her, are kept agoing exclusively by petroleum. Not a pound of meat is produced, not a head of cabbage is grown for miles around her. Every article of food consumed at Pithole has to be imported from Titusville, across miles of the worst roads in the universe." The journalist concluded with the obvious question: "How long can this supply of oil at Pithole be depended on?" The only answer he could give was that nobody knew.[19] Not exactly a vote of confidence for the industry; however, one must remember that the gamble inherent in oil speculation remained part of the attraction. To that end, a *New York Herald* correspondent described Petrolia as a place "where towns spring up in a night, as if petroleum were the genuine oil of Aladdin."[20]

Culver and the excursion planners, of course, knew the realities of the industry; however, they used such details to their ultimate advantage. While in Oil City, the group listened to a speech by Galusha A. Grow, a local business leader. During this speech, Grow assured the group that he did not believe that the supply would be exhausted "until four-inch holes were bored all over the region so close together that each would touch the other."[21] Many of the journalists took the trip's bait completely. The *New York Tribune* correspondent wrote:

> It is a misfortune that the Oil regions have too long been regarded as a theater for illegitimate and abnormal speculation, partaking more of the character of a battery than a permanent branch of mining business. The time has come when it must cease to be regarded as the pursuit of adventurers and gambling swindlers, and be governed by the same laws as all other industrial pursuits. The oil trade, in fact, is one of the heaviest and most permanent interests of the country.[22]

Simply by writing about it, these writers were helping to spread throughout the nation the model of development seen here; however, they often also gave a specific inflection to their view of economic development. Journalist E. Brooks summarized, "What is done in one small county of Pennsylvania, and upon the edge of another, may be done to a hundred counties of Virginia, Kentucky, Ohio, and elsewhere by abundant and well expended capital, as well as in far locations."[23] Many of the articles urged visitors to come and bear witness to economic progress and to take the model home to his or her own community.

The details of the place, however, began to reveal the unavoidable ills of this marriage between technology and the natural world. The mud was the single most discussed fact in each article, but there were many others. Brooks observed that he "saw not a woman, nor a cow, nor a chicken, cat, nor other domestic arrival in a ride of twenty miles."[24]

Upon observing Pithole, many journalists could not temper their words. Pithole, wrote one observer, "shows its newness most palpably. Many of the buildings are but half completed, and guiltless of paint. The new towns at the mines of California, Colorado and Montana can show nothing comparable to Pithole."[25] Such observations led some writers to make assumptions or claims about the ethics of those in the oil regions. The same correspondent went on to stress that the industry should not be allowed to proceed unregulated, particularly the drilling of too many holes. "The oil interest is too valuable to the people of the oil producing districts, and the world at large, to permit this great source of wealth to be continually tampered with by ignorant, reckless men, who rush to the oil regions intent only on making their pile in the least possible time. . . . Let not the goose that lays the golden egg be killed to gratify greedy, reckless adventurers."

"I dare say that the scenery on this creek," wrote a Boston journalist, "was once picturesque, but now it is dreary and desolate, with abandoned oil works on every hand."[26] This writer may have been the most critical observer. At every turn he saw an industrial landscape, not a social one. The towns were set up to be "perfect paradise to those who are therein enriching [their] pecuniary profit." The surroundings were ghastly to this observer. "On the creek," the oil "presents varied colors, as it is disturbed by the huge flat boats." These boats were often pulled by straining mules that walked through the water. "It was painful to witness the exertions of these poor brutes, nor did we wonder at seeing the floating remains of several that had died in harness."[27]

One writer, after observing such sights, mused at the sacred role of oil in past religions. "Oil was a sacred thing," he said, " but whatever the reverence its frequent use in sacred ceremonies may inspire, a visit to this slimy city [Oil City] must well nigh drown it forever."[28] Of Oil City, this observer judged that "of all unholy places for the habitation of mankind, this affords least promise of redemption." This observation, however, led the writer to reflect on the ethics of those residing in such a place: "A people a trifle less selfish than those of Oil City could in a single month, even with all this travel to ob-struct their operations, have roads made through their town in which the life of man or horse would not be risked at every rod."

Even these candid observations, however, could not overcome the simple fact that the selfish, exploitative tendencies were some of the most mythic and intriguing aspects of Petrolia. Another observer acquiesced that Petro-lia was "unquestionably the richest piece of country" of its kind. Its inhabi-tants, however, moved the correspondent to reflect on "how far a week's so-journ among the oil wells tends to lower one's idea of the practical greatness of the American people." This correspondent derided the American as one who revels in the "unlimited blessings which are forced upon him, takes all that comes to him and does nothing to advance his own interests."[29] The writer also delineated how much oil was produced and then how much was wasted. He estimated the annual financial loss just from the absence of roads to be $4.5 million.

Culver and the other organizers may have feared that by seeing the de-tails of the oil industry, American observers might decide that it could not be a permanent establishment—or that no one should call Petrolia home. Such fears, however, would have been unfounded. From these journalistic ac-counts, we can see that most observers thought it impossible for the indus-try and supply to last in its present form. "Every man," stated one journalist of Petrolians, "seems to act as if the sole business of his life was to make money. Thitherward all their energies are bent, and while engaged in their pursuits they scarcely give a thought to aught else. Their families are away, and are visited only once a week at the furthest, and the living is extremely simple."[30] This observer believed the priorities of oilmen to be ridiculous. The roads are so poor, he said, that transporting one barrel of oil six miles costs $3; however, if the capital were spent to repair the roads, this cost would

Fig. 7.3. Excursion illustration, from *Harper's Weekly*

drop dramatically. Simply, in Petrolia there was no interest in putting money into permanent or even semipermanent amenities. Other writers also referred to the industrialists of Petrolia as innately "selfish."

The project, however, by taking advantage of the mythic attraction of the industry and place—and even exploiting participants by subjecting them to some of its most vile discomforts—attracted a wide popular audience. *Harper's Weekly* provides the best example of this coverage, as it created no less than an advertisement for the Reno project by using three lithographs and a strip of copy, one-third of which quoted Culver about Reno (fig. 7.3).[31] The *Harper's* coverage illustrates the confidence that Petrolia could thrive forever in the form of Reno. For instance, one correspondent wrote: "Men now worked wells and lived near them keeping their families at a distance because they had no fit place to put them and bring up their children. Men came to the oil regions now to make money, get rich, and go away to enjoy the fruits of their industry. They did not intend to stay long, because they erected the commonest character of houses, and lay in the most primitive fashion. At

Fig. 7.4. General view of Petrolia with equipment, pump house and derricks

Reno all this could be remedied."[32] Another writer, after contrasting Reno with all else that he had seen in Petrolia, compared the site favorably to the idyllic scenes of the Hudson River valley.[33] These writers spread a point of view that helped to spur interest and investment in Petrolia's expansion.

The critical judgment of other writers foreshadowed Reno's fate. The excursion, in the end, tried to press the idea of permanence on an industry and a place incapable of supporting such a notion. Some visitors clearly saw that the ideals of quaint stability could not hold up under the pressures of heavy industry. While the growth of Reno and its rail line occurred swiftly, it did not compare to its decline. Within a year of the excursion, Pithole's supply of oil had run dry, and the town would disappear completely by 1867. With Pithole's demise, Reno had no present and no future. Delusions of permanence had been based on a finite resource; it was a lesson about the nature of the oil industry as well as for the planners of the region's first permanent oil town. Petrolia was an industrial site, in which one could make a fortune; but it was not a place to call home (fig. 7.4). With no supply of oil to support it, planning a comfortable city became senseless. With no reason to construct a community there, Reno's future evaporated before it had even finished being dreamed. And with Reno went Culver.

On March 28, 1866, the *New York Times* reported: "We learn definitely of the failure of Messrs. Culver, Penn & Co., Bankers of this City. Their embarrassment, we understand, arose from some railroad and other heavy undertakings. It is hoped that these embarrassments will be only temporary."[34] Culver's failure, of course, brought the failure of his banks, which led to great turmoil in the Oil Creek valley. In addition, this failure led to a rapid shift in the general opinion of the man. Two days later, six banks in the region were reported to have failed. In early April the *Titusville Herald* ran the following poem:

Reno! Reno! no words can tell
What fame you have begotten
By gulling everybody well
That you were sound, when rotten.

. . . .

Reno! Reno! don't leave us now
For you we love most dearly
And then a dry financial cow
Would starve us sure, or nearly.

Take only greenbacks for your stocks
They're Uncle Samuel's tenders.
Then with your base on pure gold rocks
Throw wildcat stock to rag-vendors.[35]

The banker crashed heavier than nearly any other figure of Petrolia, going broke, being arrested for fraud, and finally becoming one of the most infamous scourges of the oil regions. Culver was placed in prison, and Reno and its railroad would never experience the trumpeted success.

In a matter of days, Culver had gone from a hero of the region to one of its most accursed villains. His vision of Reno also evaporated, and the town never developed further than any of the other processing sites.

Despite its transitory nature and its inability to aspire otherwise, Petrolia remained of interest to Americans. The oil regions, as did many industrial sites, came to represent an ongoing relationship—at times confrontational, at others passive—between humans and the natural world. The exciting con-

frontation placed humans at the edge of danger and the industry constantly
near the possibility of extinction. Reno went against these awful realities, as
well as against the components of Petrolia's mythic landscape. By doing so,
Reno presents us with a basic truth of extractive industries: the realities of
industrial development loom too strong for visions of the technological sublime to endure for more than a moment.

The scene of early oil attracted American interest nearly as much as did
the commodity's economic potential. In essence, the temporal nature of the
towns and industry, the fires and floods, and the prostitutes and speculators
were all portions of the scene that attracted the itinerant eye's view to this
place; the rivers, hills, trees—even the oil—were merely props in this important drama going on upon the landscape of Petrolia. All became components of the transitory nature of this industry and this place as they were defined in the 1860s. To try to rewrite the drama—in other words, to construct
a Reno—simply could not succeed.

The fact that Reno exists today only as a small refining town down the Allegheny from Oil City, and not as a stable corporate, family community is
the legacy of the early industry—of Petrolia. This is the same legacy that can
be seen in the vast empty tracts of young forest along Oil Creek between Titusville and Oil City. It is a legacy that can only be perceived through a thor-
ough knowledge of this place's past.

Today, the hollow emptiness of the Oil Creek valley contrasts dramatically with the memories of the world's first oil boom. The fascination and
even awe that comes as one considers what went on here is little different
from the fascination that gripped Americans reading of the great excursion
of 1865. In fact, they are part of the same process. The remaining site offers
the viewer a statement concerning the American idea of natural abundance.
Ironically, it is a lesson that also occurred to one of the 1865 excursionists,
when this region stood as anything but empty.

> Gifted with a strange inventive genius, [the American] invents machines
> which double the amount of production with a given amount of labor, and
> then turns round and wastes in a single day all that his mechanical ability has
> gained for him in a week. We are so accustomed to finding at our very doors
> all that is needed for human happiness and human comfort, that we fairly lose
> sight of the fact that we are today throwing away in reckless profusion wealth
> enough to make us the richest people on the globe. Nowhere can a better ex-
> ample of this be found than in this oil region of Pennsylvania.[36]

I believe in the forest, and in the meadow, and in the night in which the corn grows. We require an infusion of hemlock spruce or arbor-vitae in our tea. There is a difference between eating and drinking for strength and from mere gluttony. . . . Give me wildness whose glance no civilization can endure.
—Henry David Thoreau, *Walking*

Epilogue

The Legacy of Petrolia

"People had faith in the company, that it would stay and take care of them. [Its departure] has taken away their hope that the company that started here would stay here." These words from an oil company employee could have been spoken at anytime in the late nineteenth century as populations shifted and companies closed throughout Petrolia to move on to the greener patches of the Southwest and Far West. Instead, they are contained in a 1995 *New York Times* article, which carries the headline, "Inside Oil City, Hope Runs Dry for Workers."[1]

Quaker State, the company started in Oil City and Emlenton, Pennsylvania, in 1931, had elected to move its headquarters from its new $6 million facility in Oil City to Dallas, Texas. The company had asked a few workers to move with the company, but the vast majority would become unemployed. Shocked and disappointed, these workers were simply participating in a long tradition, which the oil industry had perfected during its earliest years. The uncertainty of petroleum supplies prohibited permanence from the oil landscape of Pennsylvania. In fact, depletion and decline function as the inevitable end of any extractive economy. In essence, complete extraction is the goal pumping each derrick's swing.

These ethics were obvious from the physical landscape of the earliest stages of the industry. The users didn't care about the occupants' long-term future, just as they failed to account for that of the natural environment. But this does not mean that the scene of early oil is without meaning or significance. In fact, through illustrations such as Pithole, one finds quite the opposite. There are great stories and important lessons available from these sites, and the information is often found in the technology and land use demonstrating how residents valued this place and its commodity.

For instance, I know a place near the abandoned site of Petroleum Centre, close to two famous oil farms, the Hayes and the Egbert, where there is a thirty-foot hole in the ground that measures twelve by seventeen feet at its top. A stone foundation can be seen, and rusted iron workings litter the area. It is a hole that could have made history—could have revolutionized the young industry—but instead it became a pit for the disposal of dead animals and junk. When it was poised on the cusp of history, it was seventy-three feet deep, well on its way to the desired depth of five hundred feet. It was hoped that at this depth it would find a pool of oil and become the first oil mine in the world. It is a magical thought—a mine of oil. But in September 1865 at seventy-three feet, the danger of asphyxiation by gas became too great, and the company's funds ran low. Possible history in the making was abandoned.

But technology has a way of rising again, as if the proverbial phoenix. Developers attempted another shaft along Bull Run near Franklin in 1944. It seems a strikingly recent attempt to realize such an oddly crude-sounding idea. For some reason, it is easier to imagine that such a ridiculous idea would be pursued during the early boom years, when little was known about oil's occurrence; it seems that developers should have known better in 1944. However, as we have seen repeatedly, the value of oil has a way of warping human logic. During this attempt to mine for oil, great precautions were taken for the safety of the miners. The rock walls were shored with concrete and the men were shielded by metal plates. High costs also led to the abandoning of this project. Outgrowths of such attempts, however, spurred the horizontal mining for supplies of heavy oil, including the famed Horizontal Well built by the Wolf's Head Refining Company near Franklin.[2]

As I consider these hulking remains, I clearly identify them as "special spots" where one feels connected with past occurrences. The twentieth-century appeal of such sites must certainly be the attraction of ruins. That is, the attraction comes in viewing the remains of what was. It is in the actual viewing of the remains of places—depleted oil shafts, the thousands of abandoned wellheads that litter the forests, or the actual landscape in photographic views from the 1860s—that one faces the realization that what formerly inhabited this now "empty" site defined a moment in the early history of all industrial America.

It is striking that in a society obsessed with economic and technological progress we can also grow enraptured by a nostalgic look back—a glance to see from whence we have come before we again focus on where we would like to go. The ruins of a place and industry such as Petrolia allow historians and buffs to consider the track that progress has followed through this valley.

Yet I believe this view, like that found in an automobile's rear-view mirror, comprises only a portion of the attraction to these ruins of past industries. The actual attraction runs as a spine through each of these views, even as a portion of what intrigued nineteenth-century Americans, such as those reading of or participating in the excursion of 1865. At least a part of what attracted these people to this place was that it potentially would be gone in a blink of the eye—that from the start this place was an endangered species, poised on the edge of extinction, destined, in fact designed, to expire.

Similarly, people come today in the thousands to the sites of Petrolia's industrial heritage because the great boom, the industry of oil, did in fact vanish. And this vanishing represents something much more intrinsic to humans, whether historians or tourists. Due to the clean end point of this valley's industrial story, this place has become a representation of the dividing line between transience and permanence. Most important, within this representation, this place has come to signal an utterly basic meeting: Petrolia has become for visitors, modern or past, a view of the boundary that separates the technological human from the natural world. It suggests a lesson of limits to which the human might push the natural systems around him.

Petrolia in the 1860s served as a meeting place between these two worlds and forced land users to define a new relationship with the natural environment. This altercation at the borderline shifted the relationship in a revolutionary manner, as the human further instrumentalized the natural environment. Today, the cultural memory of the interaction mixes with the current view of the scene to press visitors to think deeply about industrial uses of the natural environment. Its appeal during the 1860s, however, had only to do with the visions of the future, not of the past.

During Petrolia's heyday, Americans' interest stemmed from thoughts of the extremes that could or might be: boom or bust. Today, it is thoughts of what has been. For most contemporary observers, this communion with the past gets little beyond the odd details and amazing stories of the boom; however, this moment of memory offers an important opportunity to draw lessons from the past that can influence our expectations of life and industry today. In essence, this moment of cultural memory—what was—can reveal ideas of what could be.

In Petrolia's history we must realize that the boom comprises only a part of the story—a tale that continues to ripple outward as the Pennsylvania Department of Environmental Resources now defines this place's use and misuse. Unlike standard history, such a story possesses no end, only chapters that have been and that are to come. As a boundary between the human and natural worlds, this valley witnessed a revolutionary encroachment by the technological human; however, even in its abandoned state, the place remains a border between these two worlds. The lessons continue and compel us to come and partake of the historical tales. Petrolia becomes a parable whose meaning only now becomes evident: for instance, that of life on the edge, brought home to Oil City—the last real oil town in Pennsylvania—as recently as 1995. The lesson is renewed again by peering down each of the abandoned oil mines, at the wellheads beneath the quiet flow of Oil Creek, or at the grave markers in the Petroleum Centre cemetery, the last remnant of the once thriving center of the valley's industry.

Throughout the history here, this unique interaction between nature and technology has been guided by the choices and priorities made by the human community that inhabited the locale. Whether choosing to sink a well into the third sand or to set aside the Oil Creek valley as a state park, these decisions are guided by the ethics and values that the culture selects as its goal or ideal. This search for meaning creates a landscape, which then can relate the moral and ethical guidelines of human users.

Petrolia was a defining point for American industrial ideas of land use and resource management. It offered an area remote from urban centers but easily accessible at a time when the nation and a large force of men were prepared to work. It came at a time when markets for illuminants and lubricants were being defined by the continued industrial revolution, and it fell in line as the most practical source for each. An energy revolution took place as petroleum and other sources of motive power became widely available. The availability allowed this revolution to completely remake human civilization and powered an industrial age like no one had imagined. Additionally, Petrolia occurred during an age of mechanization that saw nature as an instrument. The industry would move on, as they always do; what remained of the culture and the ecology had no choice but to remain.

Defined by an industry that cared only about a single commodity, the sense of this region and place became entirely one-dimensional. As the resource has dried up, the sense of the place has become only of what was—a backwards glance at the boom that came and went. Many of the towns of Petrolia have joined other extractive communities as worn-out towns whose day has passed. Petrolia was a place whose sense of itself had been defined by an external view based entirely in the commodity available there. This view dominates life here today as well.

So what is the legacy of Petrolia? I would argue that it is contained in my hyperbole above: "as they always do." Petrolia defined the oil industry, and in many senses it also helped to define the system of all extractive booms to follow. Part of the oil boom's importance is its timing, but a significant portion also derives from the details that were Petrolia.

This site's mixture of myth and fascination made the active dismantling and sacrificing of a place acceptable, if not desirable, to Americans. Industry considered planning or forethought unnecessary as it repeatedly established human communities to take finite resources out of the earth. The facts bear out Petrolia's role in this development. Petrolia introduced the practice of creating industrial wastelands to a nation with an economic drive for growth that could not be satiated. This ethic of wasting a place for the common good, of extracting a needed resource at the cost of all else in that locale, would power American industry into the modern era. Such development rarely necessitated planning or forethought to allow the ecological or cultural community also to thrive.

One can argue that portions of this legacy can be found throughout the United States, including the city of Los Angeles; the nuclear dumps of Hanford, Washington; the strip-mined stretches of West Virginia, Pennsylvania, and other states, now "reclaimed"; the Los Alamos testing site; Prince William Sound, Alaska, as well as the Alaskan pipeline region; the great, dismantled forests of the Northwest and Great Lakes that have been harvested through clear-cutting; the nuclear storage site at Yucca Mountain; the oil landscape of the Gulf of Mexico; and even in more abstract examples such as the American vision of progress and expansion. At least a portion of these manmade wastelands is Petrolia's true legacy. Each one extends out from or radiates through the Oil Creek valley or the Pithole historic site as a contemporary example of the ethic that would create wastelands in order to satiate motives that resemble those that created Petrolia.

This distinction will not be as readily claimed as was that of being the "birthplace of oil" or the "valley that changed the world." But this valley stands as the site where the technological human met the natural world under a laissez-faire system of development. Throughout its history the commodity's increasing value and, early on, its scarcity compounded these factors. The combination of these variables left the landscape of Pennsylvania forever changed. Within these changes, we find the roots of long-term shifts in ideas of natural resource management.

Today the Oil Heritage Park is a lovely natural space teeming with life and history. But the apparent natural harmony enjoyed by the visitors of today is only the latest chapter in the relationship between humans and nature in this locale. This relationship is the foundation of all that has happened here since humans arrived. In essence, the establishment of such historic sites has become the next step in the natural order. Earlier, I argued that the technological sublime had aided Americans in seeing technological change as part of a natural order of things; it appears that the present stage in this order is often the site's preservation for entirely passive uses, such as recreation. It has become the final stage, after extraction, exhaustion, and abandonment—almost as a type of apology.

Bike trails knit together the abandoned lands along today's Oil Creek valley on both sides of the creek. Small signposts crop up periodically along these paths. Each one bears a black-and-white photo from this tranquil valley's economic boom years. They record the frenzy that came as a torrent through the valley after Drake's well came in. The wooded, "empty" sight before each photo is matched up with the depicted location, presenting a striking exhibit of re-photography. The photos reveal towns, farms of derricks, teams of men and beasts, all where now there is only second-growth, mixed deciduous forest. In little more than a century this valley has made the astounding transition from industrial hub and economic boom area to a state-funded recreational site. And people still come.

The site of industry—even if it is steeped in memory—interests the itinerant eye. Once again, the explanation must lie in the technological sublime. If one accepts the idea that industrial sites can be imbued with a feeling of human progress and success, he or she must also grant that the present scene's attraction derives most from cultural memory: the idea that this massive industry was here and is gone, similar to a lost civilization. Oddly, few remnants of technology meet one's gaze; however, one must peer deeper into this story to find that the very emptiness of this place's present is a creation of extraction. The emptiness is by design.

The realization comes as we face the idea that part of what attracts us here is that the place lost its industry—its sense of worth. The complications unfold as one stands along this quiet stream and realizes just how deeply the national culture has appropriated this place's sense and meaning. The very idea of loss derives from the external view; internally, the valley's history teams with its present beauty to make it a rare treat. But just as its present state remains an example of a continuum of interaction between technology and nature, its emptiness must also be viewed for its symbolic meaning.

Here we hear echoes of David Potter and William Cronon: the people of plenty were a people of waste. And we hark back to Ida Tarbell's description of her childhood home as the site of the industry most "destructive of beauty, order, [and] decency." In this contemporary scene of beauty and tranquillity, we can consider these words and more accurately complete the vision. The scene inevitably becomes one of natural regeneration. Within the same site, the present scene shows the industry lost as well as nature's recovery. The details of the stripping and extraction of this valley's resources can no longer symbolize nature's domination by technology. The natural regeneration allows us to see that the story does not end there. Instead, the scene of the oil boom becomes more accurately one of haphazard, disorganized human futility in the shadow of nature's persistence.

Such futility stems from the repercussions of allowing greed alone to control development. In a place blessed with enormous resources, humans have found no long-term reasons to survive. We have looked no deeper than the resources we need as a reason for the existence of place. The natural environment dutifully rides out the extraction of its resources, then it covers over as if to show the ecological community's true dominance. Like the waves continuously crashing upon the shore, such reclamation presents the observer with a taste of the insignificant scale of our human actions. No matter how gross our abuses and transgressions upon the landscape, we remain a blip in geological time and merely a component of a biological ecosystem. Even here, a place more instrumentalized during the 1860s than nearly any other, the border can be taken back.

This understanding is the resource now revealed by this locale. This resource occurs within each second-growth seedling and each abandoned length of pipe buried beneath one hundred years' worth of leaves. It occurs in the abandoned sites of Petroleum Centre and Pithole, to name only two. And more than anywhere else it can be found in the flowing waters of Oil Creek. Regardless of its manipulation, abuse, or disregard, the stream has cut its path through these hills.

Such an understanding becomes a resource, similar to oil that once mined requires a refining process in which it is mixed with other substances in order to create the desired end product. The additive that complements each detail of this valley is knowledge—an awareness of what transpired there over a hundred years ago. The end product of this combination, then, is a lesson, an understanding, an ethic—that humans must at all times operate within nature.

Appendix

*Oil Boom Towns with Post Offices of Petrolia,
Listing Years of Operation*

Bedford County
    Cherry Grove (1862–70)
Butler County
    Petrolia (1872–present)
Erie County
    Corry (1861–present)
Forest County
    West Hickory (1866–present)
Indiana County
    Foster (1883–84)
Venango County
    Champion (1866)
    Columbia Farm (1869–85)
    Dennison (1865–68)
    Eagle Rock (1865–95)
    Funkville (1864–66)
    Laytonia (1864–71)
    McClintockville (1870–82)
    Miller Farm (1865–66)
    Oil City (1861–present)
    Oleopolis (1865–87)
    Petroleum Centre (1864–93)
    Pioneer (1866–89)

    Pit Hole Centre (1865–66)
    Pit Hole City (1865–95)
    Reno (1865–present)
    Rouseville (1862–present)
    Rynd Farm (1882–85)
    Shamburgh (1867–94)
    Sheridan City (1866)
    Sherman Wells (1864–72)
    Sidney (1868–77)
    South Oil City (1871–91)
    Sugar Creek (1872–94)
    Tarr Farm (1864–75)
    Wegeforth City (1866–67)
    West Hickory (1866)
    West Pithole (1866)
    Witherups (1868–87)
Warren County
    Bully Hollow (1863–64)
    Cherry Grove (1883–95)
    Cobham (1864–1904)
    Fagundus (1871–1918)
    Farnsworth (1882–84)

Notes

Introduction. The Persistence of Oil on the Brain

1. Ida Tarbell, "My Start in Life," in *All in the Day's Work: An Autobiography* (New York, 1939), 9.

2. S. J. M. Eaton, *Petroleum: A History of the Oil Region of Venango County, Pennsylvania* (Philadelphia, 1865), 42.

3. Ida Tarbell, *The History of the Standard Oil Company*, 2 vols. (New York, 1904). Excerpted in *Scribner's*, Tarbell's work was a shocking realization for many Americans who had idolized Rockefeller's rise to success as the "American way." Tarbell carefully demonstrated the unethical and, in fact, illegal use of influence by which Rockefeller had created a model of the modern corporation through vertical integration of the various processes involved in the oil industry. As a consequence, Americans questioned Rockefeller's values as well as their own.

4. Harold F. Williamson and Arnold R. Daum, *The American Petroleum Industry: The Age of Illumination, 1859–1899* (Evanston, Ill.: Northwestern University Press, 1959), 118, 737; Andrew Cone and Walter R. Johns, *Petrolia: A Brief History of the Pennsylvania Region* (New York, 1870), 618.

5. Railroads are generally considered the first example of long-distance financial speculation. Advertisements for oil companies in Pennsylvania appeared in newspapers and magazines throughout the 1860s. For more information on the image of the Pennsylvania oil boom in 1860s popular media see Paul H. Giddens, *Pennsylvania Petroleum, 1750–1872: A Documentary History* (New York: Appleton, 1970).

6. Any environmental studies textbook discusses this issue. For instance, see Andrew Goudie, *The Nature of the Environment*, 3d ed. (Cambridge: Basil Blackwell, 1993), 368. Goudie draws the landscape as a construction of the interplay between the natural world and the modifications exerted by its inhabitants. Obviously, these changes are carried out through a person's or a group's choices and priorities, which

are controlled by ethics and values. Once humans instrumentalize the natural environment and consider the application of technology in order to increase the value of the resource, they have set themselves apart from other inhabitants.

7. Eastburn, *Oil on the Brain* (Boston, 1865).

8. Anthony F. C. Wallace, *Rockdale* (New York: Norton, 1978), 4.

9. Howard Mumford Jones, *The Age of Energy* (New York: Viking Press, 1971), 104, 107.

10. Alexis de Tocqueville, "A Fortnight in the Wilds," in *Journey to America* (Westport, Conn.: Greenwood, 1981), 329.

11. Carolyn Merchant, *Ecological Revolutions* (Chapel Hill: University of North Carolina Press, 1989), 2. Merchant offers that ecological revolutions "altered the local ecology, human society, and human consciousness. . . . And . . . the forms of consciousness—perceiving, symbolizing, and analyzing—through which humans socially constructed and interpreted the natural environment were reorganized."

12. Aldo Leopold, *A Sand County Almanac, and Sketches Here and There* (New York: Oxford University Press, 1966), 205.

13. "An ethic," said Leopold, "may be regarded as a mode of guidance for meeting ecological situations so new or intricate, or involving such deferred reactions, that the path of social expediency is not discernible to the average individual" (ibid., 203).

Chapter 1. "A Good Time Coming for Whales"

1. Data from the U.S. Census Bureau reflect that the whale oil market increased in scale and scope through 1850. Massachusetts consistently led the nation by producing 3–4 million barrels of whale oil per year. New York often produced more than one million barrels, including 3.3 million in 1850. During those years, the whale oil fishery was included in the census of manufactures and industry. Since 1860, any information on whaling has been included with fishing.

2. Herman Melville, *Moby Dick* (New York: New American Library, 1980), 511.

3. Walter S. Tower, *A History of the American Whale Fishery* (Philadelphia, 1907), 66.

4. Alexander Starbuck, *History of the American Whale Fishery from Its Earliest Inception to the Year 1876* (Waltham, Mass., 1878), 8.

5. Obed Macy, *History of Nantucket* (Boston, 1835), 44–46.

6. The most helpful source for the technology and practices of whaling is Elmo Hohman, *The American Whalemen* (Clifton, N.J.: Augustus M. Kelley, 1972).

7. Bureau of the Census, *Census 1850*, Digest of Statistics of Manufactures, no. 1, Principal Manufactures, *Oil, whale.*

8. For a wonderful history of the western Arctic whaling period, see John R. Bockstoce, *Whales, Ice, and Men* (Seattle: University of Washington Press, 1986).

9. Tower, *American Whale Fishery*, 77.

10. Ibid.

11. Quoted in Hildegarde Dolson, *The Great Oildorado* (New York: Random House, 1959), 42–43.

12. The idea of commodification is part of theoretical models that derive from modern and postmodern cultural theories. The deconstruction of commodities has shown that a natural resource is assigned a value by the surrounding or using culture. If placed in a capitalist economy, this value can transfer the resource into a commodity, in which its use and management is orchestrated in a varying manner depending on its value and therefore its supply. In 1965 French anthropologist Maurice Godelier wrote, "There are thus no resources as such, but only possibilities of resources provided by nature in the context of a given society at a certain moment in its evolution." "The Object and Method of Economic Anthropology," in *Relations of Production*, ed. David Seddon (London, 1978), 61. For discussion of this topic's relation to natural resources, see William Cronon, *Changes in the Land* (New York: Hill and Wang, 1983), and *Nature's Metropolis: Chicago and the Great West* (New York: Norton, 1991).

13. Williamson and Daum, *American Petroleum Industry*, 33. Williamson and Daum's effort remains the premier work in the history of illumination and should be consulted for more details on the subjects dealt with here in brief.

14. Ibid., 35.

15. Paul H. Giddens, *The Birth of the Oil Industry* (New York, 1938), 19–20.

16. S. F. Peckham, "Production, Technology, and Uses of Petroleum and Its Products," *House Miscellaneous Documents*, no. 42, 47th Cong., 2d sess., 1883, vol. 13, pt. 10.

17. "Gas Light Companies in the United States," *American Gas-Light Journal* 4 (June 15, 1863): 373.

18. Bruce G. Trigger and Wilcomb E. Washburn, *The Cambridge History of the Native Peoples of the Americas* (New York: Cambridge University Press, 1996), 1:234–47.

19. *History of Venango County, Pennsylvania* (Columbus, Ohio, 1879), 71–72.

20. Eaton, *Petroleum*, 44–55.

21. Daniel K. Richter, *The Ordeal of the Longhouse* (Chapel Hill: University of North Carolina Press, 1992), 13–15.

22. Rev. David Zeisberger, "The Diaries of Zeisberger Relating to the First Missions in the Ohio Basin," *Ohio Archaeological and Historical Publications* 21 (n.d.): 79.

23. Gen. William Irvine, *Pennsylvania Archives*, 1st ser., 11, no. 25, 516, Drake Well Museum and Archive, Titusville, Pa.

24. Gen. Benjamin Lincoln, in *Memoirs of the American Academy of Arts and Sciences* (Boston, 1785), 1: 375.

25. P. C. Boyle, ed., *The Derrick's Hand-Book of Petroleum: A Complete Chronological and Statistical Review of Petroleum Developments* (Oil City, Pa., 1898), 1: 8.

26. Thomas Gale, *The Wonder of the Nineteenth Century: Rock Oil in Pennsylvania and Elsewhere* (Erie, Pa., 1860), 15.

27. John Earle Reynolds, *In French Creek Valley* (Meadville, Pa., 1938), 72–73.

28. Giddens, *Pennsylvania Petroleum*, 5–8.

29. Benjamin Franklin, *Autobiographical Writings*, ed. Carl Van Doren (New York, 1945), 1: 298–306. British scientists would use the same principle with petroleum in the port of Aberdeen in 1882, when they installed pipes to discharge crude directly into the ocean. Benjamin Vincent, *Haydn's Dictionary of Dates* (New York, 1889), 647.

30. Paul H. Giddens, *Early Days of Oil: A Pictorial History of the Beginnings of the Industry in Pennsylvania* (Princeton, N.J., 1948), 3–6.

31. Edwin C. Bell, *History of Petroleum* (Titusville, Pa., 1900), 150; *Pittsburgh Dispatch*, August 7, 1892.

32. See, for example, Arthur Bining, *Pennsylvania Iron Manufacture in the Eighteenth Century* (Harrisburg, Pa.: PHMC, 1973).

33. *History of Venango County*, 180.

34. See, for example, Frederick M. Binder, *Coal Age Empire: Pennsylvania Coal and Its Utilization to 1860* (Harrisburg, Pa.: PHMC, 1974).

35. Martin Melosi, *Coping with Abundance* (Philadelphia: Temple University Press, 1985), 26.

36. The fuel crisis associated with the trade blockade during the War of 1812 is often credited with establishing the market for anthracite coal. One of the greatest changes that followed this fuel crisis was an expansive drive to harvest and put to use the natural resources of the state in order to prevent periods of want from occurring again. H. Benjamin Powell, *Philadelphia's First Fuel Crisis: Jacob Cist and the Developing Market for Pennsylvania Anthracite* (University Park: Pennsylvania State University Press, 1978).

37. J. T. Henry, *The Early and Later History of Petroleum* (Philadelphia, 1873), 60–61.

38. Brewer, Account of meeting with Crosby, from *Titusville Morning Herald*, January 28, 1881, Brewer Papers, Drake Museum.

39. Benjamin Silliman, *Report on the Rock Oil, or Petroleum, from Venango County, Pennsylvania* (New Haven, Conn., 1855).

40. The story of how the first drilling proceeded is impossible to verify, as it is based on oral histories, stories, and newspaper accounts. Most important, all the facts are suspect owing to the story's dependence on the recollections of the few involved parties—especially considering each one's potential interest in taking credit for more responsibility than he might have deserved.

41. *Philadelphia Times*, September 11, 1879.

42. *Titusville Weekly Herald*, January 15, 1880.

43. Smith would report that he had done so. Interview, *Titusville Weekly Herald*, January 15, 1880.

44. Williamson and Daum, *American Petroleum Industry*, 79. James M. Townsend reports that the company had stopped supporting Drake during the summer and that Townsend alone was doing so. In early August, Townsend elected to discontinue his support and asked Drake to conclude the project. Townsend Collection, no. 412, Drake Museum.

45. Townsend Collection, no. 261, Drake Museum.

46. Interview with Samuel B. Smith, *Oil City Derrick*, August 27, 1909.

47. *Titusville Weekly Herald*, January 15, 1880.

48. *New York Tribune*, September 13, 1859.

49. *Titusville Morning Herald*, July 27, 1866.

50. Daniel Boorstin, *The Republic of Technology: Reflections on Our Future Community* (New York: Harper and Row, 1978), 24.

51. *The Living Age* (July–September 1860): 810, 812.

52. Williamson and Daum, *American Petroleum Industry*, 234.

53. The internal combustion engine would begin to use gasoline to power automobiles around 1895. Daniel Yergin, *The Prize: The Epic Quest for Oil, Money, and Power* (New York: Simon and Schuster, 1991), 79.

54. *Journal of the Franklin Institute (JFI)* 73 (1862): 375.

55. *JFI* 93 (1872): 425, 426.

56. For a discussion of the contemporary American dependence on oil, see either Melosi: *Coping with Abundance*, or Yergin, *The Prize*.

Chapter 2. *"A Triumph of Individualism"*

1. Henry, *History of Petroleum*, 92, 95–98, 108, 109.

2. Boyle, *Derrick's Hand-Book*, 1: 149.

3. John Locke, *Two Treatises on Government* (New York: Adler's Foreign Books, 1963).

4. Ibid., 343.

5. *Harper's* 50 (1865): 60.

6. See, for example, John Stilgoe, *The Common Landscape of America, 1580–1845* (New Haven: Yale University Press, 1982), and Theodore Steinberg, *Nature Incorporated: Industrialization and the Waters of New England* (New York: Cambridge University Press, 1991).

7. Ida Tarbell, introduction to Giddens, *Birth of the Oil Industry*, xxxix.

8. Garrett Hardin, "The Tragedy of the Commons," *Science* 162 (1968): 1245.

9. Williamson and Daum, *American Petroleum Industry*, 759.

10. George A. Blanchard and Edward P. Weeks, *The Law of Mines, Minerals, and Mining Rights*, 759–60. Cited in J. Stanley Clark, *The Oil Century: From the Drake Well to the Conservation Era* (Norman: University of Oklahoma Press, 1958), 97–99.

11. Ibid., 761.

12. Ibid., 760–61.

13. *Brown v. Vandergrift*, 80 Penn St. 142 (November 1875).

14. The additional cases dealing with the rule of capture took many different approaches to the issues at hand. Some sought to treat oil and gas by the laws governing the collection of wild animals or solid minerals. In actuality, though, this approach to the rule differs little from that seen in *Acton v. Blundell*. See Blakely M. Murphy, *Conservation of Oil and Gas: A Legal History* (New York: Ayer, 1972), 427. Not until the turn of the century would the rule of capture finally acquire specific legal meaning. Three Pennsylvania state supreme court cases are generally cited as authority. In these cases the right to possess all the below-ground minerals as well as the right to prevent drainage by drilling off-set wells received explicit judicial sanction. However, the ambiguous nature of the rule was all that oil speculators needed to initiate a new pattern of resource exploitation and land use. Indeed, these later rulings actually offered justification for the practices seen in the 1860s oil fields. An 1897 decision reads, "Whatever gets into the well belongs to the owner of the well no matter where it comes from." *Kelly v. Ohio Oil Company*, 49 N.E. Report 401 (December 1897).

15. Clark, *Oil Century*, 97.

16. *Westmoreland and Cambria Natural Gas Company v. Dewitt et al.*, 130 Penn St. 235 (November 1899).

17. In discussing a similar situation, legal historian Arthur McEvoy writes of the rule of capture's impact on the California fisheries. In his model, many different users with a variety of technologies harvest the common property that they all share. Such a system of jurisdiction possesses no incentives to prohibit users from overharvesting the resource. Although it was called growth, progress, prosperity, or expansion, the basic root of this desire is defined as simple human greed. Economists refer to this indulgence of human nature as "the Fisherman's Problem," which stipulates that in a competitive economy, no market mechanism ordinarily exists to reward individual forbearance in the use of shared resources. The only possible outcome of such a boom is that eventually everyone goes broke—unless they pull out and move elsewhere. As McEvoy observes, "Collectively and inevitably—tragically to [Garrett] Hardin's mind—industry degrades and eventually destroys resources owned in common but used competitively." Arthur McEvoy, *The Fisherman's Problem* (New York: Cambridge University Press, 1986), 11.

18. *Erie (Pa.) Weekly Gazette*, October 6, 1859.

19. *New York Semi-Weekly Tribune*, October 7, 1859.

20. *Coudersport (Pa.) Potter Journal*, October 13, 1859.

21. Williamson and Daum, *American Petroleum Industry*, 90–92.

22. In this practice, a ten- to fifteen-foot pole would be inclined and suspended overhead, over the hole that had been dug by hand to six or ten feet. Rope stirrups were arranged in which two men each placed a foot. These men would then

alternate kicking outward and downward, thereby also forcing the beam or pole downward.

23. Through soft shale or dirt, five to six feet could be kicked down in a day; through more difficult substances, two to three inches a day was common. Drilling three feet a day was usually considered to be a good average rate of progress. Usually, the center bit would become dull after drilling only two feet. It would then be replaced by a larger, blunter reamer. This process would require pulling the entire rig out of the ground—no matter what depth it had reached. This was also necessary to rid the hole of excess rock, water, and dirt. In the overall construction of a rig, the auger stem followed these interchangeable bits and reamers.

The auger stem was a twenty-foot-long shaft that was heavy enough to carry the force of percussion drilling. "Jars," which were two looped irons, then were attached to the stem to give it a foot of play for each stroke. These jars took the strain off the other parts of the rig. A rope socket then joined the tool string to the kicking cable, which was a rope line that then came up to the surface and was attached to the kicking beam. This work could be heartbreaking, because holes often caved in or flooded. In such cases, the rigs and tools could be lost or broken, which greatly increased the expense of sinking the well.

Once the oil flow was established, the well had to be tubed so that the supply could be controlled. This process began with lowering lengths of two- or three-inch copper piping down the hole in twelve-foot sections. Laborers then lowered bags of flaxseed to certain positions along the outside of the piping to hold it in place, and finally a pumping apparatus (usually, a wooden sucker rod with a rubber plunger attached to the end of it) directly into the tubing.

24. Eaton, *Petroleum*, 149.

25. *Venango Spectator* (Franklin, Pa.), November 5, 1862.

26. Cone and Johns, *Petrolia*, 156, 153–54.

27. Eaton, *Petroleum*, 142.

28. Deposition of Elijah Brady, September 28, 1868, at Patents Office, New York City, quoted in Williamson and Daum, *American Petroleum Industry*, 151.

29. E. Mills to Roberts, Titusville, May 20, 1865. Quoted in Williamson and Daum, *American Petroleum Industry*, 151.

30. *Scientific American* 15 (July 21, 1866): 54.

31. Williamson and Daum, *American Petroleum Industry*, 154.

32. The original American colonies are probably the nearest approximation to these zones of resource extraction that ever existed on U.S. land. For an explanation of the commodification of the landscape during the colonial period, see Cronon, *Changes in the Land*, and Timothy Silver, *A New Face on the Countryside* (New York: Cambridge University Press, 1990).

33. The irony of his situation was obvious to everyone but Drake. Ultimately, he was left destitute and very near death in New York City. When oil barons heard of

his situation, they petitioned the Commonwealth of Pennsylvania to support this man, whose "discovery" had brought them so much wealth. Drake and his family were soon put on the state's dole at $1,500 per month.

34. *Jamestown* (N.Y.) *Journal*, March 28, 1860.

35. George Bissell to his wife, November 4 and 7, 1859.

36. Cone and Johns, *Petrolia*, 222–23.

37. *Brown v. Vandergrift*, 80 Penn St. 142 (November 1875).

Chapter 3. The Sacrificial Landscape of Petrolia

1. *New York Times*, December 20, 1864, 5.

2. Cultural geography is the field most responsible for studying this intersection. Excellent sources for a broad view of the field include D. W. Meinig, ed., *The Interpretation of Ordinary Landscapes* (New York: Oxford University Press, 1979); George F. Thompson, ed., *Landscape in America* (Austin: University of Texas Press, 1995); and Michael Conzen, ed., *The Making of the American Landscape* (Boston: Unwin Hyman, 1990).

3. See Meinig, *Interpretation*, and Yi-fu Tuan, *Topophilia: A Study of Environmental Perception, Attitudes, and Values* (Englewood Cliffs, N.J.: Prentice-Hall, 1974).

4. There have been excellent contributions to the conceptual understanding of the meaning of places from a variety of fields. See, for instance, Yi-fu Tuan, *Space and Place* (Minneapolis: University of Minnesota Press, 1977), and Simon Schama, *Landscape and Memory* (New York: Vintage Press, 1996).

5. Henry, *History of Petroleum*, 337–40.

6. *Eclectic Magazine* (February 1862): 264.

7. The material for this chapter was not gathered as a cross-section but as an exhaustive analysis of all the coverage of the early oil industry within the popular media. The end product here focuses on the most indicative or suggestive articles or popular reference. Most often, these articles come from American newspapers, business journals, and upper-class general interest journals.

8. Warren I. Susman, *Culture as History: The Transformation of American Society in the Twentieth Century* (New York: Pantheon Books, 1984), 8.

9. *New York Times*, December 20, 1864.

10. *Harper's* 50 (1865): 60.

11. Ibid., 54.

12. Ibid.

13. Many of these genres conform to those laid out for the period as a whole by literary scholar David S. Reynolds, in *Beneath the American Renaissance* (Cambridge: Harvard University Press, 1989), 15.

14. *Harper's* 50, 562–63.

15. Ibid., 565.

16. *Merchants' Magazine* (July 1862): 27.

17. Ibid., 26.

18. *Continental Monthly* 5 (January 1864): 187–202.

19. *New York Times*, April 2, 1866.

20. *Harper's* 50 (1865): 569.

21. Ibid.

22. *Venango Spectator*, January 11, 1865.

23. *New York Times*, December 20, 1864.

24. This is not to suggest the unprovable point that the industry manipulated the media. Instead, the media catered to the interests of readers, which also happened to benefit investment and therefore industry.

25. *New York Times*, December 20, 1864.

26. *Merchants' Magazine* (February 1865): 6, 94.

27. *Merchants' Magazine* (May 1863): 393.

28. This article discusses the development of Canadian oil wells. *New York Times*, April 2, 1866.

29. Richard Slotkin, *The Fatal Environment: The Myth of the Frontier in the Age of Industrialization, 1800–1890* (New York: Atheneum, 1985), 52.

30. John Earle Reynolds explains that the interest in sensationalism has been part of every time period and culture; yet, the early and mid-nineteenth century "was unique since for the first time this hunger could be fed easily on a mass scale." Reynolds, *French Creek Valley*, 169.

31. Flooding in Petrolia was not covered by the national media, although it was a product of parts of the myth of Petrolia. This flooding was caused by the abuse of the landscape as oil seekers abandoned any consideration of future uses or needs of the soil and land. The stripping of forests, reckless erection of derricks and refining plants, and general lack of consideration for erosion and the region's future contributed to the flood problems.

32. Margaret H. Hazen and Robert M. Hazen, *Keepers of the Flame: The Role of Fire in American Culture, 1775–1925.* (Princeton: Princeton University Press, 1992), 8.

33. Ibid., 4.

34. The reliance of this analysis on *New York Times* coverage is a twofold necessity: indexes exist for nineteenth-century issues; and, most importantly, by picking up accounts from other newspapers, the *Times* of this period acted as an early wire service by controlling the information that would be most widely distributed.

35. *New York Times*, February 15, 1866.

36. Ibid., March 23, April 1, May 27, 1866.

37. Ibid., June 1, 1866.

38. Ibid., May 22, 1871.

39. *The Beverly Hillbillies*, which appeared in the 1960s.

40. William Wright, *The Oil Regions of Pennsylvania: Showing where Petroleum is found; How it is obtained, and at what cost* (New York, 1865), 107.

41. *Cornhill Magazine* 5 (1862): 748.

42. *New York Times*, February 10 and March 9, 1865; August 20, 1866.

43. Ibid., December 2, 1872.

44. Cone and Johns, *Petrolia*, 16–19.

Chapter 4. Oil Creek as Industrial Apparatus

1. This is common material that can be found in any history of the early industry. It was believed that the underground pools were only accessible from the lowest possible surfaces, which most geologists agreed were the lowlands along rivers. Eaton and Cone and Johns are the best sources to consult.

2. The term *instrumentalization* derives from cultural theory and anthropology. Such a process is related to commodification, but it does not necessarily involve a value in terms of potential sale. Instead, portions of the natural environment are considered singularly for their usefulness within limited industrial processes. A resource's utility within a specific system makes it into an instrument within a technological system or process.

3. Michael Williams, *Americans and Their Forests* (New York: Cambridge University Press, 1991), 82.

4. Later, the lake would be given its present name, after Chief Canadaughta of the Cornplanter people of the Six Nations. *History of Venango County*.

5. Later, a dam specifically for this purpose would be added just below Titusville.

6. At this time, seventeen sawmills and dams were located on the principal branches of Oil Creek. Giddens, *Birth of the Oil Industry*, 103.

7. Boyle, *Derrick's Hand-Book*, 1: 34.

8. Giddens, *Birth of the Oil Industry*, 104.

9. *Warren (Pa.) Mail*, January 24, 1863.

10. Ibid.

11. Ibid.

12. *Venango Spectator*, May 21, 1862.

13. *Titusville Gazette* and *Oil Creek Reporter*, June 26, 1862.

14. Andrew Carnegie, *Autobiography of Andrew Carnegie* (New York, 1920), 138.

15. Eaton, *Petroleum*, 166.

16. The term *steward* is used here to signify those making decisions guiding land use in the valley.

17. J. H. A. Bone, *Petroleum and Petroleum Wells* (Philadelphia, 1865), 79.

18. Map of Venango County, Pa., William Schuchman and Brothers, 1857; Map of Venango County, Pa., Martin and Rundall, 1865. Drake Museum.

19. These and many other details of the structures of buildings in company towns are discussed in Richard Francaviglia, *Hard Places* (Iowa City: University of Iowa Press, 1991).

20. Eaton, *Petroleum,* 197.

21. Cone and Johns, *Petrolia,* 259, 261.

22. Bone, *Petroleum and Wells,* 75.

23. This design helped to protect the structures from passing skiffs and other debris. Samuel T. Pees, "Off-Shore Drilling and Other Oil Creek Valley Innovations," Heritage Lecture Series, April 13, 1995. Bibliographic information from Pees to author, conversation.

24. Bone, *Petroleum and Wells,* 80.

25. Boyle, *Derrick's Hand-Book,* 1: 27.

26. *Venango Spectator,* May 13, 1863; *Warren Mail,* May 30, 1863.

27. Bone, *Petroleum and Wells,* 55–56, 64.

28. For a discussion of the typology carried out and suggested by such presentations, see Elizabeth Johns, *American Genre Painting* (New Haven: Yale University Press, 1991).

29. F. W. Beers, *Atlas of the Oil Regions* (New York, 1865).

Chapter 5. "What Nature Intended This Place Should Be"

1. *Venango Spectator,* April 27, 1864.

2. Beers, *Atlas,* 30.

3. Carolee Michener, *Franklin: A Place in History* (Franklin, Pa.: Franklin Bicentennial Committee, 1995), 61; quote from *Oil City Derrick,* December 1881.

4. Consider the definition of culture put forward by geographer Peter Jackson: "Culture consists of patterns, explicit and implicit, of and for behavior acquired and transmitted by symbols, constituting the distinctive achievements of human groups, including their embodiments in artifacts; the essential core of culture consists of traditional ideas and especially their attached values; culture systems may, on the one hand, be considered as products of action, on the other as conditioning elements of further action." Peter Jackson, *Maps of Meaning* (London: Unwin Hyman, 1989), 17.

5. Ian Barbour, *Technology, Environment, and Human Values* (New York: Praeger, 1980), 68.

6. Donald Worster, *Dust Bowl* (New York: Oxford University Press, 1979), 164.

7. Bureau of the Census, *Census 1870,* vol. 3, table 13, 760.

8. Additional petroleum production came from Kentucky's one establishment, Ohio's 25, and West Virginia's 140. Ibid., table 14, 769.

9. Ibid., table 15, 785–87.

10. Ibid., vol. 1, table 2, 58.

11. In 1860, 1,088 of the Venango County population of 25,043 were foreign born, and 23,955 native born. Of the total Venango County population in 1870, 42,139 were native born and 5,786 were foreign born. Ibid., table 5, 320.

12. For instance, of the native-born population, 11,601 had one or both foreign-born parents. Of these American-born individuals, 10,863 had foreign fathers and 10,033 had foreign mothers; 9,295 had both. Ibid.

13. See for instance, Terry G. Jordan and Matti Kaups, *The American Backwoods Frontier* (Baltimore: Johns Hopkins University Press, 1989).

14. Of these individuals, 2,395 had arrived from Ireland, 1,157 from British America, 955 from Germany, 855 from England and Wales, and fewer than one hundred from other European countries. This information was not available from the 1860 Census. *Census 1870*, vol. 1, table VII, 368–69. The other nationalities are Polish, Dutch, Swiss, Swedish, and French.

15. When the boom ended around 1872, it was a foregone conclusion that most laborers would move on with the industry or follow their own dreams. In 1880, the foreign-born population had fallen by 30 percent to 4,001, with a consistent drop across the nationalities: the Irish population fell to 1,654; the English and Welsh to 719; the German to 661; and the British American to 561. *Census 1880*, vol. 1, table 14, 526.

16. Ibid., table 2, 59; *Census 1860*, vol. 1, table 4, 438. These individuals were not landholders in the oil business but worked in different laboring and service portions of the oil industry. The African American population would continue to grow; by 1880, the population had risen to 547. *Census 1880*, vol. 1, table 5, 406–7.

17. Because the organization of the tabulations varied, figures needed to be estimated by compilation. The 1860 population of males ages 18 to 45 can be estimated at 5,500. The 1870 population of males 18 to 45 then skyrocketed to 12,312; the 21-and-up group was even higher at 13,665. *Census 1870*, vol. 1, table 14, 634.

18. Unfortunately, the data for cities and towns are not broken down into gender groups, so it is impossible to quantify the gender balances between boomtowns and cities.

19. Indeed, *Census 1870* places the number of women involved in the oil industry nationally at 39 oil refinery operatives and 1 well operator or laborer. None of these were located in the Pennsylvania oil regions. *Census 1870*, vol. 1, table 20, 840–41.

20. *Census 1880*, vol. 1, table 3, 318.

21. *History of Venango County*, 504–5.

22. *Titusville Courier*, March 9, 1871.

23. *History of Venango County*, 516.

24. The only occupation listed in the census that is related to the oil industry is refiner, which most likely applied only to types of oil other than petroleum (particularly kerosene). The number of refiners in Pennsylvania is listed as 50. *Census 1860*, vol. 1, "Occupations in the U.S.," 672–73.

25. *Census 1870*, vol. 1, table 18, 820.

26. While the refining of petroleum was largely conducted outside of the state, its production was still almost exclusively carried out in the state. Almost the entire workforce in both industries was made up of males aged 16 to 59 (1,654 and 3,777, respectively). Refinery operators were predominantly natives of the United States, but there also were significant numbers who were natives of Germany (350), Ireland (305), and England (130). Of well operators and laborers, 3,225 were from the United States, with another 176 coming from British America, 168 from Ireland, and 104 from Germany. Ibid., table 20, 840–41.

27. *Census 1870*, vol. 3, table 10, 614–15.

28. *Census 1860*, vol. 3, table 1, 531.

29. *Census 1870*, vol. 3, table 11, 730.

30. The operation was so efficient and its location so good that it attracted the expansive eye of the Standard Oil Company, which purchased it as part of its pioneering work in vertical integration during the early 1870s. *History of Venango County*, 510.

31. Ibid., 508.

32. *Census 1860* reported that Venango County overall possessed 57 churches, which could accommodate 18,125 members. *Census 1870* reported that the number had jumped to 89, with accommodations for 28,075. *Census 1860*, vol. 2, "Selected Statistics of Churches," 454–59; and *Census 1870*, vol. 2, table 18, 553.

33. From 1860 to 1870, however, the Methodists added only one church—a twenty-fourth—and their membership fell by 400; on the other hand, in this span Presbyterians added 15 churches to increase to 30 and to expand their capacity for membership from 5,500 to nearly 10,000.

34. Various structures were used until the Exchange dissolved in 1873. Finally, with the assistance of Rockefeller's domination of the industry, the structure still known as the Oil Exchange was dedicated in 1878, which assured that Oil City would remain an important cog in the oil business well into the twentieth century.

35. Charles A. Seely, "A Week on Oil Creek," *Scientific American* (n.d.). Reprinted in the *Titusville Morning Herald*, September 1, 1866.

36. *New York Times*, August 8, 1865.

37. *History of Venango County*, 515.

38. *New York Times*, June 1, 1866. Picked up from the *Philadelphia Evening Bulletin*.

39. *New York Times*, July 15, 1866.

40. *History of Venango County*, 513.

41. *Pittsburgh Commercial*, March 19, 1865.

42. *Oil City Register*, quoted in *History of Venango County*, 508.

43. Gale, *Wonder of the Nineteenth Century*, 33.

44. The only early oil town not located in Venango County, Titusville is just over the Crawford County line. In fact, during the oil boom there was a short-lived move-

ment to create a new county based on the oil regions, mainly to bring Titusville into the fold with the rest (and possibly to have been called Petrolia).

45. In 1753, Major George Washington had been dispatched from Virginia to northwestern Pennsylvania in order to inform the French troops and residents who were living in Fort Machault, present-day Franklin, to leave the territory that now belonged to England. He traveled throughout the area but spent little time in the immediate vicinity of the fort. This area would then be involved in the French and Indian War, and the fort would eventually become a British frontier outpost, Fort Venango, and later Fort Franklin.

46. Quoted in Michener, *Franklin*, 43.

47. *Census 1860*, vol. 3, table 1, 504; *Census 1870*, vol. 3, table 11, 724.

48. Michener, *Franklin*, 45.

49. *Titusville Morning Herald*, August 12, 1865.

50. Ibid., August 18, 1865.

51. Ibid., August 23, 1865.

52. Giddens, *Early Days*, 98.

53. Yergin, *The Prize*, 43.

54. See ibid., 42–46, for a description of these events.

55. Ibid., 56.

Chapter 6. Pithole: Boomtowns and the "Drawing Board City"

1. The well, by this time, was more a sentimental interest than a financial producer.

2. U.S. Petroleum Company *(First Annual) Report*, New York, December 1, 1865, 4–5.

3. *Titusville Morning Herald*, June 29, 1870.

4. J. B. Jackson, *Discovering the Vernacular Landscape* (New Haven: Yale University Press, 1984), 12. The natural environment bears little pertinence in Jackson's landscape hierarchy unless it is set off by human boundaries for some cultural reason such as preservation or conservation.

5. Petition for incorporation, September 1, 1865. Drake Archives.

6. William Culp Darrah, *Pithole, The Vanished City: A Story of the Early Days of the Petroleum Industry* (Gettysburg, Pa.: author, 1972), iv.

7. J. J. Bouton, *New York Journal of Commerce*, October 20, 1865.

8. Alfred Brunson, *A Western Pioneer* (1872), 215, cited in Darrah, *Pithole*, 7.

9. *Oil City Register*, November 24, 1864.

10. Lease, Thomas Holmden to James Faulkner Jr., Venango County Deed Book Z, 19, recorded May 2, 1864.

11. Darrah, *Pithole*, 10.

12. *Oil City Register*, January 8, 1865.

13. Darrah, *Pithole*, 16.

14. Ibid., 17.

15. *New York Tribune*, August 18, 1865.

16. Darrah, *Pithole*, 21.

17. *Titusville Morning Herald*, August 16, 1864; *Pithole City Directory*, 1865–66, 6.

18. *Oil City Register*, June 1, 1865.

19. *New York Herald*, July 30, 1865.

20. *Titusville Morning Herald*, July 29, 1865.

21. *New York Herald*, July 20, 1865.

22. C. C. Leonard, *History of Pithole* (Pithole, Pa., 1867), 47.

23. Darrah, *Pithole*, 146.

24. *Nation* 1 (1865): 371.

25. Ibid.

26. There were very few carpenters in the area, so most construction was done by the proprietors.

27. Darrah, *Pithole*, 38.

28. Ibid., 61.

29. *Titusville Morning Herald*, July 25, August 2, 1865; *Oil City Register*, August 17, 1865.

30. *Titusville Morning Herald*, September 14, 1865.

31. Leonard, *History of Pithole*, 16.

32. *Titusville Morning Herald*, December 7, 1865.

33. *Pithole Daily Record*, December 14, 1865.

34. Darrah, *Pithole*, 23.

35. S. Morton Peto, *Resources and Prospects of America, 1866* (New York, 1866), 184.

36. *Philadelphia Press*, August 8, October 10, 1865.

37. J. N. Fradenbaugh, *History of the Erie Conference* (Oil City, Pa., 1907), 2: 353–60.

38. *Titusville Morning Herald*, July 29, 1865.

39. John J. McLaurin, *Sketches in Crude Oil: Some Accidents and Incidents of the Petroleum Development in All Parts of the Globe* (Harrisburg, Pa., 1898), 169–70.

40. *Pithole Daily Record*, December 1, 1866, April 4–5, 1867.

41. Ibid., January 6, 1866.

42. McLaurin, *Sketches in Crude Oil*, 142.

43. Darrah, *Pithole*, 111.

44. *Pithole Daily Record*, throughout early January 1866.

45. Ibid., April 4, August 3, 1866.

46. Jackson, *Discovering the Vernacular Landscape*, 40.

Chapter 7. Delusions of Permanence

1. *New York Times*, October 28, 1865.

2. For a discussion of this concept, see, for instance, Yi-fu Tuan, *Space and Place*.

3. "Oil Excursion," *Weekly Toledo Blade*, October 21, 1865.

4. *New York Times*, "An Excursion to the Oil Regions," October 25, 1865.

5. Tuan, *Topophilia*, 63.

6. Paul Shepard, *Man in the Landscape* (New York: Knopf, 1967), 156.

7. Ibid., 127.

8. This message was often subsumed within a different motive: "the sublime was being absorbed into a religious, moral, and frequently nationalistic concept of nature, contributing to the rhetorical screen under which the aggressive conquest of the country could be accomplished." Barbara Novak, *Nature and Culture* (New York: Oxford University Press, 1980), 38.

9. Novak stresses that this change marked the demythologizing of the landscape, the making human of nature (163).

10. David Nye, *The American Technological Sublime* (Cambridge: MIT Press, 1996), 111.

11. John F. Kasson, *Civilizing the Machine* (New York: Grossman, 1976), 174.

12. Kasson writes that such a scene combines a realistic mode with a "sense of mythic grandeur that intensifies the sublime atmosphere while containing it within a positive symbolism. . . . The total effect is a compelling and ultimately reassuring tribute to the wonder of American technology" (170–71).

13. Actually, Culver organized two such excursions, the first of which consisted entirely of European visitors.

14. *Evening Post* (n.p.), October 21, 1865.

15. *Meadville Republican*, "The Excursion," October 18, 1865.

16. See, for instance, *Detroit Advertiser and Tribune*, "A Trip to the Oil Regions," October 23, 1865.

17. Ibid.

18. *New York Express*, October 20, 1865.

19. *Journal of American Commerce*, October 20, 1865.

20. *New York Herald*, October 23, 1865.

21. *New York Commercial Advertiser*, "From the Oil Regions," October 21, 1865.

22. *New York Tribune*, "The Petroleum Fields," October 24, 1865.

23. *New York Express*, October 23, 1865.

24. Ibid.

25. *Rochester Express*, October 23, 1865.

26. *Boston Journal*, October 21, 1865.

27. Ibid., October 23, 1865.

28. *New York Daily Times*, October 28, 1865.

29. *New York World*, October 24, 1865.

30. *New York Commercial Advertiser*, October 26, 1865.

31. *Harper's Weekly*, November 25, 1865, 740–42.

32. *New York Commercial Advertiser*, October 24, 1865.

33. *Boston Daily Advertiser*, October 24, 1865.

34. *New York Times*, March 28, 1866.
35. "Reno Rhinoed," poem, *Titusville Herald*, April 1866.
36. *New York World*, October 24, 1865.

Epilogue. The Legacy of Petrolia

1. *New York Times*, July 26, 1995.
2. This background information is taken from the *Titusville Herald*, February 16, 1987.

Select Bibliography

American Petroleum Institute. *History of Petroleum Engineering.* New York: API, 1961.

Anderson, Robert O. *Fundamentals of the Petroleum Industry.* Norman: University of Oklahoma Press, 1984.

Asbury, Herbert. *The Golden Flood: An Informal History of America's First Oil Field.* New York: Knopf, 1942.

Ashburner, Charles A. "The Bradford Oil District of Pennsylvania." *AIME Transactions* 7 (1879): 316–28.

———. "The Product and Exhaustion of the Oil-Regions of Pennsylvania and New York." *AIME Transactions* 14 (1885): 419–28.

Barbour, Ian G. *Technology, Environment, and Human Values.* New York: Praeger, 1980.

Beaton, Kendall. "Dr. Gesner's Kerosene: The Start of American Oil Refining." *Business History Review* 29 (March 1955): 28–53.

Beers, F. W. *Atlas of the Oil Regions.* New York, 1865.

Belasco, Warren James. *Americans on the Road: From Autocamp to Motel, 1910–1945.* Cambridge: MIT Press, 1979.

Bell, Edwin C. *History of Petroleum.* Titusville, Pa.: *The Bugle* Print, 1900.

Bentley, Jerome Thomas. "The Effects of Standard Oil's Vertical Integration into Transportation on the Structure and Performance of the American Petroleum Industry, 1872–1884." Ph.D. diss., University of Pittsburgh, 1976.

Bining, Arthur. *Pennsylvania Iron Manufacture in the Eighteenth Century.* Harrisburg, Pa.: PHMC, 1973.

Boatright, Mody C., and William A. Owens. *Tales from the Derrick Floor: A People's History of the Oil Industry.* Garden City, N.Y.: Doubleday, 1970.

Bone, J. H. A. *Petroleum and Petroleum Wells.* Philadelphia, 1865.

Boorstin, Daniel J. *The National Experience.* Vol. 2 of *The Americans.* New York: Vintage Books, 1965.

———. *The Republic of Technology: Reflections on Our Future Community.* New York: Harper and Row, 1978.

Boyle, P. C., ed. *The Derrick's Hand-Book of Petroleum: A Complete Chronological and Statistical Review of Petroleum Developments.* 2 vols. Oil City, Pa., 1898, 1900.

Brady, Kathleen. *Ida Tarbell: Portrait of a Muckraker.* New York: Putnam, Seaview, 1984.

Brantly, J. E. *History of Oil Well Drilling.* Houston: Gulf, 1971.

Carl, John F. *The Geology of the Oil Regions of Warren, Venango, Clarion, and Butler Counties.* Report 3. Harrisburg, Pa., 1880.

———. *Geological Report on Warren County and the Neighboring Oil Regions.* Harrisburg, Pa., 1883.

Carnegie, Andrew. *Autobiography of Andrew Carnegie.* New York: Houghton Mifflin, 1920.

Catalogs, Equipment:
   Oil Well Supply Co., Oil City, Pa., 1892, 1902, 1916.
   National Supply Co., Toledo, Ohio, 1906.

Chernow, Ron. *Titan: The Life of John D. Rockefeller, Sr.* New York: Random House, 1998.

Clark, James A. *The Chronological History of the Petroleum and Natural Gas Industries.* Houston: Clark Book Co., 1963.

Clark, J. Stanley. *The Oil Century: From the Drake Well to the Conservation Era.* Norman: University of Oklahoma Press, 1958.

Cone, Andrew, and Walter R. Johns. *Petrolia: A Brief History of the Pennsylvania Region.* New York, 1870.

Conzen, Michael, ed. *The Making of the American Landscape.* Boston: Unwin Hyman, 1990.

Cronon, William. *Changes in the Land.* New York: Hill and Wang, 1983.

———. *Nature's Metropolis: Chicago and the Great West.* New York: Norton, 1991.

Crum, A. R., ed. *The Romance of American Petroleum and Gas.* 2 vols. Oil City, Pa., 1911.

Darrah, William Culp. *Pithole, the Vanished City: A Story of the Early Days of the Petroleum Industry.* Gettysburg, Pa.: author, 1972.

Dolson, Hildegard. *The Great Oildorado.* New York: Random House, 1959.

Drake Well Museum and Archive, oral history recordings. Collected by David Weber, Pleasantville, Pa., these recordings are excellent cultural and social sources for the early- and mid-twentieth-century oil industry in northwestern Pennsylvania. Titusville, Pa.

Eaton, S. J. M. *Petroleum: A History of the Oil Region of Venango County, Pennsylvania.* Philadelphia, 1865.

Fanning, Leonard M. *Men, Money, and Oil.* New York: World, 1966.

Francaviglia, Richard. *Hard Places.* Iowa City: University of Iowa Press, 1991.

Franks, Kenny A., and Paul F. Lambert. *Early Louisiana and Arkansas Oil.* College Station: Texas A&M University Press, 1982.

Fursenko, A. A. *The Battle for Oil: The Economics and Politics of International Corporate Conflict over Petroleum, 1860–1930*. Greenwich, Conn.: JAI Press, 1990.

Gale, Thomas. *The Wonder of the Nineteenth Century: Rock Oil in Pennsylvania and Elsewhere*. Erie, Pa., 1860.

Geertz, Clifford. *The Interpretation of Cultures*. New York: Basic Books, 1973.

Giddens, Paul H. *The Birth of the Oil Industry*. New York: Macmillan, 1938.

———. *Early Days of Oil: A Pictorial History of the Beginnings of the Industry in Pennsylvania*. Princeton: Princeton University Press, 1948.

——— *Pennsylvania Petroleum, 1750–1872: A Documentary History*. New York: Appleton, 1970.

Giebelhaus, August W. *Business and Government in the Oil Industry: A Case Study of Sun Oil, 1876–1945*. Greenwich, Conn.: JAI Press, 1980.

Gordon, Robert, and Patrick Malone. *The Texture of Industry: An Archaeological View of the Industrialization of North America*. New York: Oxford University Press, 1994.

Gulliford, Andrew. *Boomtown Blues: Colorado Oil Shale, 1885–1985*. Niwot: University Press of Colorado, 1989.

Harvard Business School. *Oil's First Century*. Special issue, *Business History Review*. Cambridge: Harvard Business School, 1960.

Hazen, Margaret Hindle, and Robert M. Hazen. *Keepers of the Flame: The Role of Fire in American Culture, 1775–1925*. Princeton: Princeton University Press, 1992.

Henry, J. T. *The Early and Later History of Petroleum*. Philadelphia, 1873.

History of Venango County, Pennsylvania. Columbus, Ohio, 1879.

House, Boyce. *Oil Boom*. Caldwell, Idaho: Caxton Printers, 1941.

Jackson, John Brinckerhoff. *Landscapes: Selected Writings of J. B. Jackson*. Edited by Ervin H. Zube. Amherst: University of Massachusetts Press, 1970.

———. *The Necessity of Ruins*. Amherst: University of Massachusetts Press, 1980.

———. *Discovering the Vernacular Landscape*. New Haven: Yale University Press, 1984.

———. *A Sense of Place, a Sense of Time*. New Haven: Yale University Press, 1994.

Jackson, Peter. *Maps of Meaning*. London: Unwin Hyman, 1989.

Johns, Elizabeth. *American Genre Painting*. New Haven: Yale University Press, 1991.

Johnson, Arthur Menzies. *The Development of American Petroleum Pipelines: A Study in Private Enterprise and Public Policy, 1862–1906*. Ithaca: Cornell University Press, 1956.

Jones, Howard Mumford. *The Age of Energy*. New York: Viking Press, 1971.

Jordan, Terry G., and Matti Kaups. *The American Backwoods Frontier*. Baltimore: Johns Hopkins University Press, 1989.

Kasson, John F. *Civilizing the Machine*. New York: Grossman, 1976.

Knowles, Ruth Sheldon. *The Greatest Gamblers: The Epic of American Oil Exploration*. New York: McGraw-Hill, 1959.

Lambert, Paul F., and Kenny A. Franks. *Voices from the Oil Fields.* Norman: University of Oklahoma Press, 1984.

Leonard, C. C. *History of Pithole.* Pithole, Pa., 1867.

Leopold, Aldo. *A Sand County Almanac, and Sketches Here and There.* New York: Oxford University Press, 1966.

———. *The River of the Mother of God and Other Essays.* Madison: University of Wisconsin Press, 1991.

Limerick, Patricia Nelson. *The Legacy of Conquest: The Unbroken Past of the American West.* New York: Norton, 1987.

Lynd, Robert S., and Helen Merrell Lynd. *Middletown.* New York: Harcourt, Brace, 1929.

McEvoy, Arthur. *The Fisherman's Problem.* New York: Cambridge University Press, 1986.

McLaurin, John J. *Sketches in Crude Oil: Some Accidents and Incidents of Petroleum Development in All Parts of the Globe.* 2d ed. Harrisburg, Pa., 1898.

McPhee, John. *The Pine Barrens.* New York: Farrar, Straus and Giroux, 1968.

McWilliams, Carey. *Factories in the Field.* Boston: Little, Brown, 1939.

Malamud, Gary W. *Boomtown Communities.* New York: Van Nostrand Reinhold, 1984.

Mancall, Peter C. *Valley of Opportunity.* Ithaca: Cornell University Press, 1991.

Marsh, George P. *The Earth as Modified by Human Action.* New York, 1874.

Marx, Leo. *The Machine in the Garden.* New York: Oxford University Press, 1964.

Meinig, D. W., ed. *The Interpretation of Ordinary Landscapes.* New York: Oxford University Press, 1979.

Melosi, Martin. *Coping with Abundance.* Philadelphia: Temple University Press, 1985.

Merchant, Carolyn. *Ecological Revolutions.* Chapel Hill: University of North Carolina Press, 1989.

Michener, Carolee. *Franklin: A Place in History.* Franklin, Pa.: Franklin Bicentennial Committee, 1995.

Miller, Ernest C. *Oil Mania: Sketches from the Early Pennsylvania Oil Fields.* Philadelphia: Dorrance, 1941.

———. *This Was Early Oil: Contemporary Accounts of the Growing Petroleum Industry, 1848–1885.* Harrisburg, Pa.: Pennsylvania Historical and Museum Commission, 1968.

———. *Pennsylvania's Oil Industry.* Gettysburg, Pa.: Pennsylvania Historical Association, 1974.

Mumford, Lewis. *The Culture of Cities.* New York: Harcourt, Brace, 1938.

Murphy, Blakely M. *Conservation of Oil and Gas: A Legal History.* New York: Ayer, 1972.

Norton, William. *Explorations in the Understanding of Landscape.* Westport, Conn.: Greenwood Press, 1989.

Novak, Barbara. *Nature and Culture*. New York: Oxford University Press, 1980.

Nye, David. *American Technological Sublime*. Cambridge: MIT Press, 1996.

The Oil City Derrick's Statistical Abstract of the Petroleum Industry. Oil City, Pa., 1916.

Oil's First Century: Papers Given at the Centennial Seminar on the History of the Petroleum Industry. Cambridge: The Harvard Graduate School of Business Administration, 1959.

Olien, Roger M., and Diana Davids Olien. *Oil Booms*. Lincoln: University of Nebraska Press, 1982.

Pratt, Joseph A. *The Growth of a Refining Region*. Greenwich, Conn.: JAI Press, 1980.

Reynolds, David S. *Beneath the American Renaissance*. Cambridge: Harvard University Press, 1989.

Reynolds, John Earle. *In French Creek Valley*. Meadville, Pa., 1938.

Richter, Daniel K. *The Ordeal of the Longhouse*. Chapel Hill: University of North Carolina Press, 1992.

Roberts, Peter. *Anthracite Coal Communities*. New York: Macmillan, 1904.

Ross, Phillip W. "The Turbulent Rise of Practical Petroleum Geology, 1859–1889." Manuscript, 1992.

Ross, Victor. *The Evolution of the Oil Industry*. Garden City, N.Y.: Doubleday, Page, 1920.

Rundell, Walter, Jr. *Early Texas Oil*. College Station: Texas A&M University Press, 1977.

Schlereth, Thomas J. *Material Culture: A Research Guide*. Lawrence: University Press of Kansas, 1985.

———. *Cultural History and Material Culture: Everyday Life, Landscapes, Museums*. Charlottesville: University Press of Virginia, 1992.

Shepard, Paul. *Man in the Landscape*. New York: Knopf, 1967.

Silliman, B., Jr. *Report on the Rock Oil, or Petroleum, from Venango County, Pennsylvania*. New Haven, Conn., 1855.

Silver, Timothy. *A New Face on the Countryside*. New York: Cambridge University Press, 1990.

Simmons, I. G. *Changing the Face of the Earth*. London: Basil Blackwell, 1989.

Sisler, J. D. *Contributions to the Oil and Gas Geology of Western Pennsylvania*. Bulletin M19. Topographic and Geologic Survey. Harrisburg, Pa., 1933.

Slotkin, Richard. *The Fatal Environment: The Myth of the Frontier in the Age of Industrialization, 1800–1890*. New York: Atheneum, 1985.

Smith, Duane A. *Mining America*. Lawrence: University Press of Kansas, 1987.

Steinberg, Theodore. *Nature Incorporated: Industrialization and the Waters of New England*. New York: Cambridge University Press, 1991.

Stilgoe, John. *The Common Landscape of America, 1580–1845*. New Haven: Yale University Press, 1982.

Stocking, George Ward. *The Oil Industry and the Competitive System: A Study in Waste.* New York: Houghton, Mifflin, 1925.

Susman, Warren I. *Culture as History: The Transformation of American Society in the Twentieth Century.* New York: Pantheon Books, 1984.

Tarbell, Ida. *The History of the Standard Oil Company.* 2 vols. New York: McClure, Phillips, 1904.

———. *All in the Day's Work: An Autobiography.* New York: Macmillan, 1939.

Thomas, William L., Jr., ed. *Man's Role in Changing the Face of the Earth.* Vol. 1. Chicago: University of Chicago Press, 1972.

Thompson, George F., ed. *Landscape in America.* Austin: University of Texas Press, 1995.

Tocqueville, Alexis de. *Journey to America.* Westport, Conn.: Greenwood Press, 1981.

Townshend, Henry H. *New Haven and the First Oil Well.* New Haven: Yale University Press, 1934.

Trachtenberg, Alan. *The Incorporation of America.* New York: Hill and Wang, 1982.

———. *Reading American Photographs.* New York: Hill and Wang, 1989.

Trigger, Bruce G., and Wilcomb E. Washburn. *The Cambridge History of the Native Peoples of the Americas.* Vol. 1. New York: Cambridge University Press, 1996.

Tuan, Yi-fu. *Topophilia: A Study of Environmental Perception, Attitudes, and Values.* Englewood Cliffs, N.J.: Prentice-Hall, 1974.

———. *Space and Place.* Minneapolis: University of Minnesota Press, 1977.

Turner, Victor. *The Anthropology of Performance.* New York: PAJ, 1988.

Wallace, Anthony F. C. *Rockdale.* New York: Norton, 1978.

Watts, Mary Theilgaard. *Reading the Landscape of America.* New York: Macmillan, 1975.

White, Richard. *Land Use, Environment, and Social Change: The Shaping of Island County, Washington.* Seattle: University of Washington Press, 1995.

———. *The Organic Machine.* New York: Hill and Wang, 1995.

Whiteshot, Charles A. *The Oil-Well Driller: A History of the World's Greatest Enterprise, the Petroleum Industry.* Mannington, W.Va., 1905.

Williams, Michael. *Americans and Their Forests.* New York: Cambridge University Press, 1991.

Williamson, Harold F., and Arnold R. Daum. *The American Petroleum Industry: The Age of Illumination, 1859–1899.* Evanston, Ill.: Northwestern University Press, 1959.

Winner, Langdon. *The Whale and the Reactor: A Search for Limits in an Age of High Technology.* Chicago: University of Chicago Press, 1986.

Wolbert, George S., Jr. *American Pipe Lines: Their Industrial Structure, Economic Status, and Legal Implications.* Norman: University of Oklahoma Press, 1952.

Worster, Donald. *Dust Bowl*. New York: Oxford University Press, 1979.

————, ed. *The Ends of the Earth*. New York: Cambridge University Press, 1988.

————. *Nature's Economy*. New York: Cambridge University Press, 1991.

————. *The Wealth of Nature*. New York: Oxford University Press, 1993.

Wright, William. *The Oil Regions of Pennsylvania: Showing where Petroleum is found; How it is obtained, and at what cost. With hints for whom it may concern*. New York, 1865.

Yergin, Daniel. *The Prize: The Epic Quest for Oil, Money, and Power*. New York: Simon and Schuster, 1991.

Zimmerman, Erich W. *Conservation in the Production of Petroleum*. New Haven: Yale University Press, 1957.

Zube, Ervin H., and Margaret J. Zube. *Changing Rural Landscapes*. Amherst: University of Massachusetts Press, 1977.

Index

abandonment: of towns and leases, 149; of region, 190–97. *See also* speculation
Acton v. Blundell, 43–44
advertisements, 72
agriculture, 34, 65
Allegheny Mountains, 1, 61
Allegheny National Forest, 1
Allegheny River: fires on, 102; regional primacy of, 82–83, 124–25, 138, 181; transportation via, 86, 89–90, 164, 189
American Journal of Science and Arts, 72
American Universal Geography, 25
American West, 74, 114, 146–47
Angier, J. D., 28
anthracite coal, 26, 112. *See also* coal
Astor House. *See* Pithole

banks, first, 157. *See also* Culver, Charles Vernon
Barnsdall Well, 58
Beers, F. W., *Atlas*, 104–5
Beverly Hillbillies, The, 78–79
Bissell, George, 28–29, 58
bituminous coal, 20, 26, 112. *See also* coal
blacksmith, 48

Bliss Opera House. *See* Titusville
Blood Farm, 107–8, 111, 137
boarding houses, 56, 61, 96, 125, 163. *See also* Pithole; transience
boardwalks, 133, 159, 164
Bone, J. H. A., *Petroleum and Petroleum Wells*, 103–5
boom, 40, 44. *See also* transience
Boom Town, 78
boomtowns: community, 146–47, 150, 159–61, 164, 169–70; development, 59, 124–26, 159–61; ethnic composition of, 116–17; and gender, 116–17, 160; image of, 70, 110; law and order in, 154; Oil Creek, 95–96; religion in, 161–63; Rockefeller's attitude toward, 138; Titusville, 132–34, 136; town speculation and, 141–42. *See also* brothels; Cherrytree Run; Cornplanter; Petroleum Centre; Pithole; Red Hot; transience
Boorstin, Daniel, *The Americans*, 34
Booth, John Wilkes, 155
Bradford County, 138
Brewer, Dr. Francis, 28–29, 32, 38, 45
brothels, 157. *See also* Pithole
Brown, Thomas H., 149
Burchfield, Jim, 18

"kicking down," 48
Kier, Samuel, 25–26, 29

laissez-faire approach to development, 39, 44, 61, 137. *See also* transience
landscape: appeal of mythic, 79, 189; awful details of Petrolia, 71, 81; components of, 61; cultural, as defined by industry and corporate life, 139; meaning of boomtown, 146–47; of Petrolia as landmark trust in technology, 81; remaking by economic boom, 83; sacred, 170–71; sacrificial, 74; significance for historical inquiry, 9; sympathetic fallacy, 18; in users' ethics and values, 108, 193
lease, 54–56. *See also* speculation
Leopold, Aldo, 9
Lincoln, Gen. Benjamin, 23
Living Age, The, 35
Locke, John, and value in land, 39
long-distance financial speculation: and abandonment, 149; commodification of Oil Creek valley, 108; first extractive industry to draw from, 57; lack of stewardship interest, 132; marketers of land, 18, 55; and post–Civil War boom, 179; and reduction of risk as stimulating boom, 54; and relationship to myth, 70; use of maps, media, and guidebooks for, 66, 72, 103–5. *See also* Beers, F. W.; commodification
lubrication, 16, 27, 35
lumber, 26, 28, 34, 86, 131. *See also* energy

McClintock Farm, 25, 58, 96, 97. *See also* Petroleum Centre
Marx, Leo, 178
Mather, John: as "Oil Creek Artist," 83–85; ethics of land use revealed in

photos, 91, 97–98, 169; images contrast with lithographs, 68
mechanization. *See* industry
media coverage: of myth of Petrolia, 64–68, 70–74, 76–81; nationalism in, 67, 69; nostalgia in, 69, 70; travelogue description of Oil City, 65–66
medicinal use of oil, 23, 25, 125
Melville, Herman, 15, 16
Merchant, Carolyn, 9
Merchants' Magazine, 35, 69, 72–73
Merrimack River, 84
Midland, Texas, 142. *See also* boomtowns
Miller Farm, 164. *See also* pipelines
Moravian missionary. *See* exploration
Morey, A. G., 158–59
Morse, Jedediah, *American Universal Geography* (1789), 25
mud, 104, 122, 133, 159–60, 184. *See also* deforestation; erosion
Murphy's Theater. *See* Pithole
music of oil boom, 7, 70–71
myth: conflict and danger in, 74, 78; horrific detail in, 80; potential of, 147; Sussman's definition of, 64–65
myth of Petrolia: attitude toward safety and, 77–78, 121; components of, 71; danger in shipping of crude, 80; and flowing wells on, 52–53; impact on town development, 130; importance to economic development, 58, 71–74; increases regional exploitation, 61–62, 83; laissez-faire practices in, 39, 41–42; media coverage and, 64–68, 70–74, 76–81; potential for fortune, 35, 78–79; photographic preservation of, 84; and priorities of technological innovation, 53; role of fire and danger, 75–79, 81, 104, 173, 189; scenery of, 177, 184. *See also* danger; fire; floods; mud

About the Author

Brian Black, a recipient of the American Society for Environmental History's Aldo Leopold Prize, is a social and cultural historian who specializes in the North American landscape and environmental issues. He is the author of *America at War* (Scholastic, 1992) and has served on the editorial staffs of various journals and magazines, including *American Studies* and the *Journal of Economic History*. A native of central Pennsylvania, Black teaches history and environmental studies at Altoona College of the Pennsylvania State University.

Library of Congress-in-Publication Data
Black, Brian, 1966–
    Petrolia : the landscape of America's first oil boom / Brian Black.
       p.   cm. — (Creating the North American landscape)
    Includes bibliographical references and index.
    ISBN 0-8018-6317-1 (alk. paper)
    1. Petroleum—Pennsylvania—Oil Creek Valley (Crawford County and Venango
County, Pa.)—History.   2. Petroleum industry and trade—Pennsylvania—Oil
Creek Valley (Crawford County and Venango County, Pa.)—History.   3. Oil Creek
Valley (Crawford County and Venango County, Pa.)—Environmental conditions—
History.   I. Title.   II. Series.
    TN872.P4 B56 2000
    338.4'76223382'0974897 21—dc21   99-042473